EAI/Springer Innovations in Communication and Computing

Series editor
Imrich Chlamtac, European Alliance for Innovation, Ghent, Belgium

T0172266

Editor's Note

The impact of information technologies is creating a new world yet not fully understood. The extent and speed of economic, life style and social changes already perceived in everyday life is hard to estimate without understanding the technological driving forces behind it. This series presents contributed volumes featuring the latest research and development in the various information engineering technologies that play a key role in this process.

The range of topics, focusing primarily on communications and computing engineering include, but are not limited to, wireless networks; mobile communication; design and learning; gaming; interaction; e-health and pervasive healthcare; energy management; smart grids; internet of things; cognitive radio networks; computation; cloud computing; ubiquitous connectivity, and in mode general smart living, smart cities, Internet of Things and more. The series publishes a combination of expanded papers selected from hosted and sponsored European Alliance for Innovation (EAI) conferences that present cutting edge, global research as well as provide new perspectives on traditional related engineering fields. This content, complemented with open calls for contribution of book titles and individual chapters, together maintain Springer's and EAI's high standards of academic excellence. The audience for the books consists of researchers, industry professionals, advanced level students as well as practitioners in related fields of activity include information and communication specialists, security experts, economists, urban planners, doctors, and in general representatives in all those walks of life affected ad contributing to the information revolution.

About EAI

EAI is a grassroots member organization initiated through cooperation between businesses, public, private and government organizations to address the global challenges of Europe's future competitiveness and link the European Research community with its counterparts around the globe. EAI reaches out to hundreds of thousands of individual subscribers on all continents and collaborates with an institutional member base including Fortune 500 companies, government organizations, and educational institutions, provide a free research and innovation platform.

Through its open free membership model EAI promotes a new research and innovation culture based on collaboration, connectivity and recognition of excellence by community.

More information about this series at http://www.springer.com/series/15427

Kolla Bhanu Prakash
G. R. Kanagachidambaresan

Editors

Programming with TensorFlow

Solution for Edge Computing Applications

Editors
Kolla Bhanu Prakash
KL Deemed to be University
Vijayawada, AP, India

G. R. Kanagachidambaresan
Department of CSE
Vel Tech Rangarajan Dr Sagunthala R&D
Institute of Science and Technology
Chennai, Tamil Nadu, India

ISSN 2522-8595　　　　　　ISSN 2522-8609　(electronic)
EAI/Springer Innovations in Communication and Computing
ISBN 978-3-030-57079-8　　　ISBN 978-3-030-57077-4　(eBook)
https://doi.org/10.1007/978-3-030-57077-4

This Springer imprint is published by the registered company Springer Nature Switzerland AG
The registered company address is: Gewerbestrasse 11, 6330 Cham, Switzerland

Preface

Machine learning and deep learning approaches have become inevitable solutions in all domains of engineering. Python is an efficient tool that satisfies the needs of engineers, mathematicians, and data scientists in solving their daily problems. The algorithms like Neural Network (NN), Support Vector Machine (SVM), Hidden Markov Model (HMM) in machine learning approach create easy way for predicting the data set and aids in classifying heterogeneous data. The Python programming language has easy packages available on open source that can immediately implement and test these algorithms for real-time problems. Python is becoming a very efficient tool, capable of running on small machines (i.e. embedded systems and single-board computers) as well as big super computers and gigantic data clusters. More and more IoT-based projects and rapid prototyping are done nowadays to solve numerous transient problems using Python and ML approaches. Creating a solution to this and sophisticating the new learning of Python and Tensorflow, this book covers beginner to advanced levels. It contains 12 parts, starting with basic pip installation of packages in Linux and Windows environment, through image processing, sentimental analysis, handwriting recognition, factor analysis, feature extraction, line recognition, various machine learning approaches, single-board computers and IoT projects using Tensorflow. This book provides detailed coding explanation along with the output to facilitate the readers and aids easy learning of Tensorflow package in Python. It also provides example exercises and their solutions in the appendix. This book will be an insight for new beginners, students, scholars, and data scientists to learn and work on Tensorflow and similar packages.

Vijayawada, AP, India
Chennai, Tamil Nadu, India

Kolla Bhanu Prakash
G. R. Kanagachidambaresan

Acknowledgement

Our sincere thanks to Almighty and our parents for the their blessings, guidance, love and support in all stages of life. We are thankful to our beloved family members for standing by us throughout our career and also helping us advance our careers through editing this book.

Our special thanks to Sri. Koneru Satyanarayana, President, K. L. University, India and Vel Tech Rangarajan Dr Sagunthala R&D Institute of Science and Technology for their continuous support and encouragement throughout the preparation of this book. I dedicate this book to them.

Our great thanks to our students and family members who put in their time and effort to support and contribute in some manner. We would like to express our gratitude to all those who supported, shared, talked things over, read, wrote, offered comments, allowed us to quote their remarks and assisted with editing, proofreading and designing through this book journey. We pay our sincere thanks to the open data set providers.

We believe that the team of authors provides the perfect blend of knowledge and skills that went into authoring this book. We thank each of the authors for devoting their time, patience, perseverance and effort towards this book; we think that it will be a great asset to all researchers in this field!

We are grateful to Eliska, Mary James and all other members of Springer's publishing team who showed us the ropes to creating this book. Without that knowledge, we would not have ventured into starting this book, which ultimately led to this. Their trust in us, their guidance and providing the necessary time and resources gave us the freedom to manage this book.

Last but not least, we'd like to thank our readers, who gave us their trust, and we hope our work inspires and guides them.

Kolla Bhanu Prakash
G. R. Kanagachidambaresan

Contents

Introduction to Tensorflow Package

Kolla Bhanu Prakash, Adarsha Ruwali, and G. R. Kanagachidambaresan

Tensorflow's structure is based on the dataflow graph [3]. A dataflow graph as two basic computation units:

- Node
- Edge

A node represents any mathematical operations and an edge [4] depicts the multidimensional array (tensors). Figure 1 elucidates the graphical flow for x*y + 2 and b + W*x equations.

Figure 2 illustrates the computation of Tensorflow in multidimensional.

1   Why Tensorflow for Deep Learning?

Tensorflow has built-in supports for deep learning [5]. This helps in simplification and has an easy-to-use environment to assemble neural networks assigning parameters, training and testing any deep learning models [6]. Likewise, simple trainable mathematical functions are also present in Tensorflow. All these flexible tools within Tensorflow make itself compatible for many concepts under machine learning.

K. B. Prakash · A. Ruwali
KL Deemed to be University, Vijayawada, AP, India

G. R. Kanagachidambaresan (✉)
Vel Tech Rangarajan Dr Sagunthala R&D Institute of Science and Technology,
Chennai, Tamil Nadu, India

© Springer Nature Switzerland AG 2021
K. B. Prakash, G. R. Kanagachidambaresan (eds.), *Programming with
TensorFlow*, EAI/Springer Innovations in Communication and Computing,
https://doi.org/10.1007/978-3-030-57077-4_1

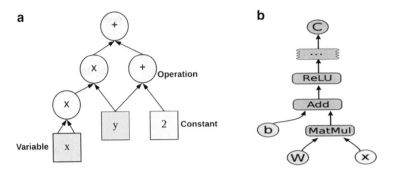

Fig. 1 (**a**) Computation Tensorflow graph for x*y + 2 (**b**) Computation Tensorflow graph for b + W*x

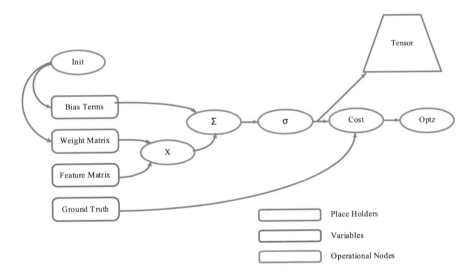

Fig. 2 Computation Tensorflow for multi-dimensional operations

2 Installation Guide to Tensorflow

2.1 System Requirement

Ubuntu 16.04 or later (64-bit)
macOS 0.12.6 (High Sierra) or later (64 bit) no GPU support
Windows 7 or later (64 bit) (python 3 only)
Raspbian 9.0 or later

If you are using python with 3.x version and have installed 'pip', then you can directly head towards the below installation in command prompt:

$pip3 install Tensorflow

Now if you have Tensorflow already installed in your [6] system and need an upgrade:

$pip3 install –ignore-installed –upgrade Tensorflow==1.9
1.9 is the version of Tensorflow to upgrade to

Following is the illustration of how python and Tensorflow both are installed in Windows 10 (64 bit):

1. Download python from https://www.python.org.downloads/ as illustrated in Fig. 3.
2. Once the python is downloaded, install it adding to the path shown as given in Fig. 4.

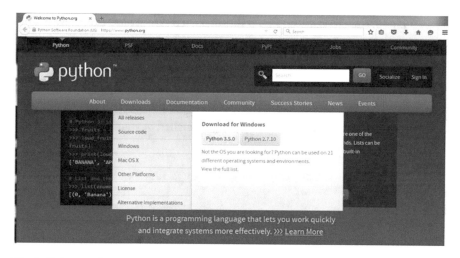

Fig. 3 Python.org home page

Fig. 4 Python download

Fig. 5 After running the above command output

3. Now to install Tensorflow as a library package in the python. Type the command shown below:

$ pip3 install Tensorflow, Fig. 5 illustrates the screen after executing the pip install command.

The version of Tensorflow installed in the existing system can be verified with the following code.

import Tensorflow as tf

tf.version

tf.version will give you the version of package installed in your system

Great! It's done.

References

1. Learning TensorFlow [Authors: Tom Hope, Yehezkel S. Resheff & Itay Lieder]
2. Deep Learning Pipeline: Building A Deep Learning Model With TensorFlow [Authors: Hisham El-Amir, Mahmoud Hamdy]
3. TensorFlow for Machine Intelligence_ A Hands-On Introduction to Learning Algorithms [Authors: Sam Abrahams, Danijar Hafner, Erik Erwitt, Ariel Scarpinelli]
4. Python Machine Learning: Machine Learning and Deep Learning with Python, scikit-learn, and TensorFlow [Authors: Sebastian Raschka, Vahid Mirjalili]
5. Python Deep Learning: Exploring deep learning techniques, neural network architectures and GANs with PyTorch, Keras and TensorFlow. [Authors: Ivan Vasilev, Daniel Slater, Gianmario Spacagna, Peter Roelants, Valentino Zocca]
6. S. Pichai, "TensorFlow: smarter machine learning for everyone", Google Official Blog, 2015.

Tensorflow Basics

Abhilash Kumar Jha, Adarsha Ruwali, Kolla Bhanu Prakash, and G. R. Kanagachidambaresan

1 Hello Tensorflow Program

This is the very basic [2] program in tensorflow and outputs are given in Figs. 1 and 2.

- *tf.constant* adds value to a given variable and the value remains constant throughout runtime.
- *tf.Session* runs a computational graph and starts the session where the objects are executed.

2 Representation of Vector/Matrix

The generalized form of vectors and matrices are called tensors [3]. There are different ways of representing tensors declaring its variability or as a constant. To represent a vector or simply to print any vector, numpy is another library package that handles the vectors and even multidimensional arrays as well. The output of each step is illustrated in Fig. 3.

A. K. Jha · A. Ruwali · K. B. Prakash (✉)
KL Deemed to be University, Vijayawada, AP, India

G. R. Kanagachidambaresan
Vel Tech Rangarajan Dr Sagunthala R&D Institute of Science and Technology,
Chennai, Tamil Nadu, India

© Springer Nature Switzerland AG 2021
K. B. Prakash, G. R. Kanagachidambaresan (eds.), *Programming with TensorFlow*, EAI/Springer Innovations in Communication and Computing,
https://doi.org/10.1007/978-3-030-57077-4_2

5

```
import tensorflow as tf

h=tf.constant("Hello ")
t=tf.constant("Tensorflow!")

ht= h+t

s=tf.Session()
print(s.run(ht))

b'Hello Tensorflow!'
```

Fig. 1 System output of simple hello program in tensorflow

```
import tensorflow as tf

h=tf.constant("Hello Tensorflow!")

s=tf.Session()
print(s.run(h))

b'Hello Tensorflow!'
```

Fig. 2 Output for HelloTensorflow basic code

```
import tensorflow as tf
import numpy as np

mat_1 = [[10,20,30],
         [40,50,60]]


mat_2 = tf.constant([[10,20,30],[40,50,60]])

mat_3 = np.array([[10,20,30],[40,50,60]])

print(mat_1)

[[10, 20, 30], [40, 50, 60]]

print(mat_2)

Tensor("Const_8:0", shape=(2, 3), dtype=int32)

print(mat_3)

[[10 20 30]
 [40 50 60]]
```

Fig. 3 Vector representation in numpy package

3   With Session() Vs without Session()

Session() helps in the execution of the computational graph and also controls the state of the tensorflow runtime [4]. Without session, tensorflow programs cannot be executed. Following are the demos of the tensorflow program with and without Session() as given in Figs. 4 and 5.

4   Zeros Matrix and Ones Matrix

Tensorflow built-in functions tf.zeros() and tf.ones() are the functions for matrix with all zeros and with all ones, respectively [5]. In the following demo, Figs. 6 and 7, the size of the matrix is 50×50 for both ones and zeros.

5   Make Matrix Negative

The following demonstration in Fig. 8 helps in negative matrix initialization, that is, matrix with –ive elements in row and [6] column initiation using the built-in function of tensorflow.

```
import tensorflow as tf

a=tf.constant([1,2,3])

print(a)
Tensor("Const:0", shape=(3,), dtype=int32)
```

Fig. 4 Without session in tensorflow

```
import tensorflow as tf

a=tf.constant([1,2,3])

s=tf.Session()

print(s.run(a))
[1 2 3]
```

Fig. 5 With session initialization in tensorflow

```
import tensorflow as tf

a=tf.ones([50,50])

s=tf.Session()

print(s.run(a))
[[1. 1. 1. ... 1. 1. 1.]
 [1. 1. 1. ... 1. 1. 1.]
 [1. 1. 1. ... 1. 1. 1.]
 ...
 [1. 1. 1. ... 1. 1. 1.]
 [1. 1. 1. ... 1. 1. 1.]
 [1. 1. 1. ... 1. 1. 1.]]
```

Fig. 6 tf.ones() example code with session

```
b=tf.ones([50,50])

print(s.run(b))
[[1. 1. 1. ... 1. 1. 1.]
 [1. 1. 1. ... 1. 1. 1.]
 [1. 1. 1. ... 1. 1. 1.]
 ...
 [1. 1. 1. ... 1. 1. 1.]
 [1. 1. 1. ... 1. 1. 1.]
 [1. 1. 1. ... 1. 1. 1.]]
```

Fig. 7 tf.ones() example code without session

```
import tensorflow as tf

a=tf.constant([2,5,6])

s=tf.Session()

print(s.run(tf.negative(a)))
[-2 -5 -6]
```

Fig. 8 Negative matrix initialization

- tf.constant, initialize the matrix with 1 × 3 dimension and store it in "a."
- To make matrix negative, tf.negative() is used.

6 Variables and Constants

The demonstration in Fig. 9 illustrates the constant and variable initializing. Here:

- Variables, constants, and placeholders are used to store values at different instances.
- Constant values cannot be altered once initialized during runtime.
- Variables assigned can be altered if required.
- Placeholders reserve space for the variables declared before runtime session.
- Global_variables_initializer() returns a variable list that holds the global variables and variable_initializer() takes all the variable in the variable list and sets its variable initializer property to an operation group [7].

7 Variables Concept in Tensorflow

- Variable holds data that vary.
- Variable holds and updates parameters in a training model.
- It is a tensor having in-memory buffers [8].
- Variables are initialized and can be stored to disk during and after training.
- That saved data can be restored to exercise the model. Figure 10 elucidates the variable concept output with Python.

```
import tensorflow as tf

x=tf.constant(35,name='x')

y=tf.Variable(x+10,name='y')

z=tf.global_variables_initializer()

sess=tf.Session()

sess.run(z)

print(sess.run(y))
45
```

Fig. 9 Code illustration for global variable initializer

```
In [1]: import tensorflow as tf

In [2]: a = tf.placeholder(tf.int32, shape=(3,1))
        b = tf.placeholder(tf.int32, shape=(1,3))

In [3]: c = tf.matmul(a,b)

In [4]: sess = tf.Session()

In [7]: print("Matrix Multiplication a,b:\n",sess.run(c,feed_dict={a:[[3],[2],[1]], b:[[1,2,3]]}))

        Matrix Multiplication a,b:
         [[3 6 9]
         [2 4 6]
         [1 2 3]]

In [ ]:
```

Fig. 10 Variable concept output

Note
A constant's value is stored in the graph and its value is repeated everywhere the graph is loaded. Separately one variable is stored.

8   Implement Concept of Placeholder

- Placeholder holds the variables and its property before, during, and after the session is executed.
- Placeholder provides the value later when the session is being executed.
- feed_dict is to feed the input value to the variable while evaluating the graph.
- feed_dict can be initialized during session runtime depicting values to variables in placeholder.

9   Simple Equations Using Tensorflow

Basic mathematical operations are used to implement an equation shown in Figs. 11, 12, and 13. Constants, variables, or placeholders can be used to form the equation. feed_dict is used to feed the [9] input values to the variable held by placeholder during session run time.

(a). Implement a + b.

```
import tensorflow as tf

a=tf.placeholder(tf.int32)
b=tf.placeholder(tf.int32)

c= tf.add(a,b)

sess=tf.Session()

print("a+b: \n",sess.run(c,feed_dict={a:5, b:5}))

a+b:
 10
```

Fig. 11 Tensorflow addition concept

```
import tensorflow as tf

a=tf.placeholder(tf.float32)
b=tf.placeholder(tf.float32)
c=tf.constant(2.0)

x=tf.multiply(a,a)
y=tf.multiply(b,b)
xy=tf.multiply(c,tf.multiply(a,b))
eq=tf.add(x,tf.add(xy,y))

sess=tf.Session()

print("Quadratic equation: \n",sess.run(eq,feed_dict={a:2, b:5}))

Quadratic equation:
 49.0
```

Fig. 12 Quadratic example program in tensorflow

(b). Quadratic equation using tensorflow.
(c). Implement $ax^2 + cxy + by^2$.

10   Simple Operations in Tensorflow

So far, we have discussed about basic operations and equations implemented in tensorflow [10]. Here are some more simple operations where tensorflow's built-in functions are used as given in Fig. 14.

```
import tensorflow as tf

a=tf.constant(5)
b=tf.constant(5)
c=tf.constant(10)
x=tf.placeholder(tf.int32)
y=tf.placeholder(tf.int32)

ax2=tf.multiply(a,tf.multiply(x,x))
by2=tf.multiply(b,tf.multiply(y,y))
cxy=tf.multiply(c,tf.multiply(x,y))
eq=tf.add(ax2,tf.add(cxy,by2))

sess=tf.Session()

print("Equation: \n",sess.run(eq,feed_dict={x:5, y:5}))

Equation:
 500
```

Fig. 13 Equation concept in tensorflow

```
import tensorflow as tf

a=tf.constant(5)
b=tf.constant(5)
c=tf.constant(25)

sums = tf.add(a,b)
sub = tf.subtract(a,b)
multi = tf.multiply(a,b)
power = tf.pow(a,b)

sess=tf.Session()

print("Sum: ",sess.run(sums))
print("Subtraction: ",sess.run(sub))
print("Multiplication: ",sess.run(multi))
print("Power: ",sess.run(power))

Sum:   10
Subtraction:   0
Multiplication:   25
Power:   3125
```

Fig. 14 Math operations in tensorflow

tf.add(x, y): add two tensors.

tf.subtract(x, y): tensors of same type are subtracted.

tf.multiply(x, y): element-wise multiplication.

tf.pow(x, y): power of x to y.

tf.exp.(x): equivalent to exponential to the power x.

tf.sqrt(x): equivalent to the power of 0.5.

tf.div(x, y): divided x by y.

tf.mod(x, y): equivalent to x%y.

References

1. Learning TensorFlow [Authors: Tom Hope, Yehezkel S. Resheff&Itay Lieder]
2. Deep Learning Pipeline: Building A Deep Learning Model With TensorFlow [Authors: Hisham El-Amir, Mahmoud Hamdy]
3. TensorFlow for Machine Intelligence_ A Hands-On Introduction to Learning Algorithms [Authors: Sam Abrahams, DanijarHafner, Erik Erwitt, Ariel Scarpinelli]
4. Python Machine Learning: Machine Learning and Deep Learning with Python, scikit-learn, and TensorFlow [Authors: Sebastian Raschka, VahidMirjalili]
5. Python Deep Learning: Exploring deep learning techniques, neural network architectures and GANs with PyTorch, Keras and TensorFlow. [Authors: Ivan Vasilev, Daniel Slater, GianmarioSpacagna, Peter Roelants, Valentino Zocca]
6. Practical Computer Vision Applications Using Deep Learning with CNNs: With Detailed Examples in Python Using TensorFlow and Kivy. [Author: Ahmed Fawzy Gad]
7. Hands-On Machine Learning with Scikit-Learn and TensorFlow: Concepts, Tools, and Techniques to Build Intelligent Systems. [Author: AurélienGéron]
8. Learn TensorFlow 2.0: Implement Machine Learning And Deep Learning Models With Python. [Authors: Pramod Singh, Avinash Manure]
9. Agrawal A, Roy K (2019) Mimicking leaky-integrate-fire spiking neuron using automotion of domain walls for energy-efficient brain-inspired computing. IEEE Trans Magn 55(1):1–7
10. Akinaga H, Shima H (2010) Resistive random access memory (reram) based on metal oxides. Proc IEEE 98(12):2237–2251

Visualizations

G. R. Kanagachidambaresan and G. Manohar Vinoothna

1 Matplotlib in Tensorflow

Pyplot provides the necessary figures and axes to get the desired plot [4]. Similarly, pylab combines the pyplot functionality with numpy as an effective environment for visualization [5].

Implementations

(a) **Scatterplot**

   Figure 1 describes the plotting of scatter plot using matplotlib package in Python.
   Before implementing further, let us get familiar with some basic terminologies used:

- **Linspace**

Syntax Used

Numpy.linspace(start, stop, num = 50, endpoint = True, retstep = False, dtype = None)

Start: start of interval (default value is 0)
Stop: end of interval
num: number of sample generated
dtype: data type of an output array

   Linspace is used to create a number sequences [6]. Implementation of this function and plotting it in a line graph is shown in Fig. 2.

G. R. Kanagachidambaresan (✉)
Vel Tech Rangarajan Dr Sagunthala R&D Institute of Science and Technology, Chennai, Tamil Nadu, India

G. Manohar Vinoothna
KL Deemed to be University, Vijayawada, AP, India

© Springer Nature Switzerland AG 2021
K. B. Prakash, G. R. Kanagachidambaresan (eds.), *Programming with TensorFlow*, EAI/Springer Innovations in Communication and Computing, https://doi.org/10.1007/978-3-030-57077-4_3

```
import matplotlib.pyplot as plt

x = [10,20,30,40,50]
y = [-1,0,1,2,3]

plt.scatter(x,y)
plt.show()
```

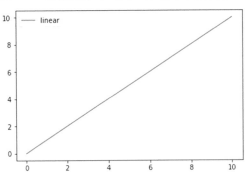

Fig. 1 Scatter plot

```
import matplotlib.pyplot as plt
import numpy as np

x = np.linspace(0,10,100)

plt.plot(x,x,label='linear')
plt.legend()
plt.show()
```

Fig. 2 Linear graph

- **np.random.randn()**

Syntax

numpy.random.randn(d0, d1, ..., dn), creates a specified array with random values w.r.t standard normal distribution as given in Fig. 3.

The values generated by this function are random [7] every time it is executed.

Using the above functions, let us do a scatter plot (Fig. 4), which contains 100 different values within the range of −10 to +10.

```
import numpy as np
```

```
x = np.random.randn(5)
```

```
print(x)
[ 0.         0.1010101  0.2020202  0.3030303  0.4040404  0.50505051
  0.60606061 0.70707071 0.80808081 0.90909091 1.01010101 1.11111111
  1.21212121 1.31313131 1.41414141 1.51515152 1.61616162 1.71717172
  1.81818182 1.91919192 2.02020202 2.12121212 2.22222222 2.32323232
  2.42424242 2.52525253 2.62626263 2.72727273 2.82828283 2.92929293
  3.03030303 3.13131313 3.23232323 3.33333333 3.43434343 3.53535354
  3.63636364 3.73737374 3.83838384 3.93939394 4.04040404 4.14141414
  4.24242424 4.34343434 4.44444444 4.54545455 4.64646465 4.74747475
  4.84848485 4.94949495 5.05050505 5.15151515 5.25252525 5.35353535
  5.45454545 5.55555556 5.65656566 5.75757576 5.85858586 5.95959596
  6.06060606 6.16161616 6.26262626 6.36363636 6.46464646 6.56565657
  6.66666667 6.76767677 6.86868687 6.96969697 7.07070707 7.17171717
  7.27272727 7.37373737 7.47474747 7.57575758 7.67676768 7.77777778
  7.87878788 7.97979798 8.08080808 8.18181818 8.28282828 8.38383838
  8.48484848 8.58585859 8.68686869 8.78787879 8.88888889 8.98989899
  9.09090909 9.19191919 9.29292929 9.39393939 9.49494949 9.5959596
  9.6969697  9.7979798  9.8989899  10.        ]
```

Fig. 3 Random number matrix generation

```
import matplotlib.pyplot as plt
import numpy as np
```

```
x = np.linspace(-10,10,100)
y=0.5 * np.random.randn(*x.shape)
```

```
plt.scatter(x,y)
plt.show()
```

Fig. 4 Linspace with scatter plot

1.1 Histogram Implementation

Syntax

plt.hist(variable used,bins = <constant value>)

X-axis denotes each value and Y-axis or the height of each [8] rectangle represents the frequency of that value. The demonstration is given in Fig. 5.

1.2 Trigonometric Curves

Matplotlib with numpy is an effective way to represent the data [9]. Trigonometric curves of sine, cosine, and tan are shown in Figs. 6, 7, and 8.

Similarly, matplotlib is also used to plot 3D figures [10]. Figure 9 is one sample of 3D graph.

```python
import matplotlib.pyplot as plt
import numpy as np


x = np.random.randn(500)


plt.hist(x, bins=100)
```
```
(array([ 1.,   0.,   0.,   0.,   1.,   0.,   1.,   1.,   1.,   1.,   0.,   1.,   4.,
         1.,   2.,   0.,   1.,   2.,   1.,   2.,   6.,   4.,   3.,   1.,   3.,   3.,
         3.,   3.,   4.,   8.,   5.,  10.,   6.,   6.,   7.,   8.,   9.,  12.,   8.,
         8.,  12.,   9.,   9.,  17.,  13.,  11.,   8.,   7.,  14.,  15.,   9.,  12.,
         7.,   9.,  12.,   8.,  13.,   9.,   7.,  12.,   8.,   7.,  10.,   9.,  14.,
        12.,  13.,   6.,   4.,   4.,   6.,   9.,   4.,   4.,   6.,   3.,   3.,   3.,
         3.,   2.,   1.,   3.,   2.,   1.,   2.,   4.,   1.,   0.,   1.,   1.,   0.,
         0.,   1.,   1.,   0.,   0.,   1.,   0.,   0.,   1.]),
 array([-2.94118882, -2.88146205, -2.82173528, -2.76200851, -2.70228174,
        -2.64255497, -2.58282821, -2.52310144, -2.46337467, -2.4036479 ,
        -2.34392113, -2.28419437, -2.2244676 , -2.16474083, -2.10501406,
        -2.04528729, -1.98556053, -1.92583376, -1.86610699, -1.80638022,
        -1.74665345, -1.68692669, -1.62719992, -1.56747315, -1.50774638,
        -1.44801961, -1.38829285, -1.32856608, -1.26883931, -1.20911254,
        -1.14938577, -1.08965901, -1.02993224, -0.97020547, -0.9104787 ,
        -0.85075193, -0.79102516, -0.7312984 , -0.67157163, -0.61184486,
        -0.55211809, -0.49239132, -0.43266456, -0.37293779, -0.31321102,
        -0.25348425, -0.19375748, -0.13403072, -0.07430395, -0.01457718,
         0.04514959,  0.10487636,  0.16460312,  0.22432989,  0.28405666,
         0.34378343,  0.4035102 ,  0.46323696,  0.52296373,  0.5826905 ,
         0.64241727,  0.70214404,  0.7618708 ,  0.82159757,  0.88132434,
         0.94105111,  1.00077788,  1.06050465,  1.12023141,  1.17995818,
         1.23968495,  1.29941172,  1.35913849,  1.41886525,  1.47859202,
         1.53831879,  1.59804556,  1.65777233,  1.71749909,  1.77722586,

         1.83695263,  1.8966794 ,  1.95640617,  2.01613293,  2.0758597 ,
         2.13558647,  2.19531324,  2.25504001,  2.31476677,  2.37449354,
         2.43422031,  2.49394708,  2.55367385,  2.61340061,  2.67312738,
         2.73285415,  2.79258092,  2.85230769,  2.91203446,  2.97176122,
         3.03148799]),
 <a list of 100 Patch objects>)
```

Fig. 5 Histogram

```
import matplotlib.pyplot as plt,plt1,p2
import numpy as np
```

```
x = np.linspace(-10,10,20)
```

```
y1=np.sin(x)
y2=np.cos(x)
y3=np.tan(x)
```

```
plt.plot(x,y1,label='Sine Curves')
plt.legend()
plt.show()
```

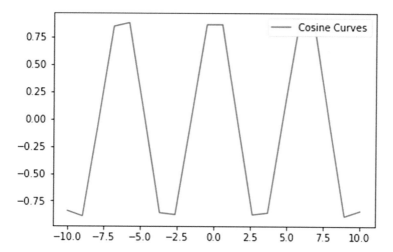

Fig. 6 Sine curve

Fig. 7 Cosine curve

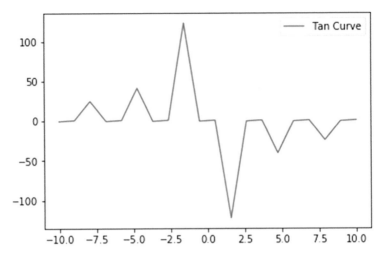

Fig. 8 Tan curve

```
from mpl_toolkits import mplot3d
import matplotlib.pyplot as plt
import numpy as np
```

```
x1=plt.axes(projection='3d')
```

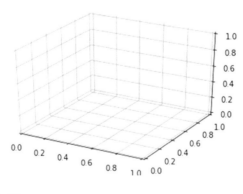

Fig. 9 3D axes

References

1. Learning TensorFlow [Authors: Tom Hope, Yehezkel S. Resheff&Itay Lieder]
2. Deep Learning Pipeline: Building A Deep Learning Model With TensorFlow [Authors: Hisham El-Amir, Mahmoud Hamdy]
3. TensorFlow for Machine Intelligence_ A Hands-On Introduction to Learning Algorithms [Authors: Sam Abrahams, DanijarHafner, Erik Erwitt, Ariel Scarpinelli]
4. Python Machine Learning: Machine Learning and Deep Learning with Python, scikit-learn, and TensorFlow [Authors: Sebastian Raschka, VahidMirjalili]
5. Python Deep Learning: Exploring deep learning techniques, neural network architectures and GANs with PyTorch, Keras and TensorFlow. [Authors: Ivan Vasilev, Daniel Slater, GianmarioSpacagna, Peter Roelants, Valentino Zocca]
6. M. Abadi, A. Agarwal, P. Barham, E. Brevdo, Z. Chen, C. Citro, G. S. Corrado, A. Davis, J. Dean, M. Devin, S. Ghemawat, I. Goodfellow, A. Harp, G. Irving, M. Isard, Y. Jia, R. Jozefowicz, L. Kaiser, M. Kudlur, J. Levenberg, D. Mane, R. Monga, S. Moore, D. Murray, C. Olah, M. Schuster, J. Shlens, B. Steiner, I. Sutskever, K. Talwar, P. Tucker, V. Vanhoucke, V. Vasudevan, F. Viegas, O. Vinyals, P. Warden, M. Wattenberg, M. Wicke, Y. Yu, and X. Zheng. TensorFlow: Large-scale machine learning on heterogeneous systems, 2015. Software available from tensorflow.org
7. D. Archambault, T. Munzner, and D. Auber. Grouseflocks: Steerable exploration of graph hierarchy space. Visualization and Computer Graphics, IEEE Transactions on, 14(4):900–913, 2008.
8. J. Abello, F. Van Ham, and N. Krishnan. ASK-Graphview: A large scale graph visualization sytem. Visualization and Computer Graphics, IEEE Transactions on, 12(5):669–676, 2006.
9. Hands-On Deep Learning for Images with TensorFlow: Build intelligent computer vision applications using TensorFlow and Keras [Authors: Will Ballard]
10. Deep Learning with Applications Using Python: Chatbots and Face, Object, and Speech Recognition with Tensorflow and Keras. [Authors: Navin Kumar Manaswi]

Regression

Kolla Bhanu Prakash, Adarsha Ruwali, and G. R. Kanagachidambaresan

1 Regression Model – Simple Linear Equation

Following are the steps that show the sample for linear equation and its plot (Fig. 1).

1. Import the numpy library and matplotlib.pyplot library.
2. Define number of variables necessary for the program.
3. Iterate the variables for 500 random points.
4. Plot the generated points using matplotlib (Fig. 2).

2 Linear Regression

Linear regression [4] [5] models predict the correlation among 2 variables or factors as shown in Fig. 3. The factor that is being guessed is called dependent variables and the factors [6] that were used to guess the value of dependent variables are called independent variables [7].

The equation for linear regression is:

$$Y = MX + C$$

K. B. Prakash (✉) · A. Ruwali
KL Deemed to be University, Vijayawada, AP, India

G. R. Kanagachidambaresan
Vel Tech Rangarajan Dr Sagunthala R&D Institute of Science and Technology,
Chennai, Tamil Nadu, India

© Springer Nature Switzerland AG 2021
K. B. Prakash, G. R. Kanagachidambaresan (eds.), *Programming with TensorFlow*, EAI/Springer Innovations in Communication and Computing,
https://doi.org/10.1007/978-3-030-57077-4_4

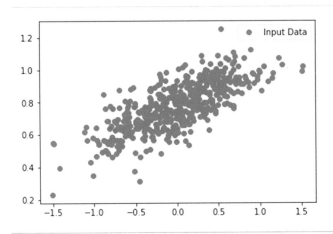

Fig. 1 Linear equation plot with matplotlib

```
import matplotlib.pyplot as plt
import numpy as np

number_of_points = 500
x_point = []
y_point = []
a = 0.22
b = 0.78

for i in range(number_of_points):
    x = np.random.normal(0.0,0.5)
    y = a*x+b+np.random.normal(0.0,0.1)
    x_point.append(x)
    y_point.append(y)

plt.plot(x_point,y_point,'o',label='Input Data')
plt.legend()
plt.show()
```

Fig. 2 Steps and code snippet using matplotlib

Here, M = Gradient of line

X, Y = co-ordinate axes
C = Y-intercept

Algorithm Used
Functionalities of each [8] cell shown in the program below are given in the following steps:

1. Import tensorflow, numpy, and matplotlib.pyplot.
2. Initialize total number of epochs, learning rate, and display step.

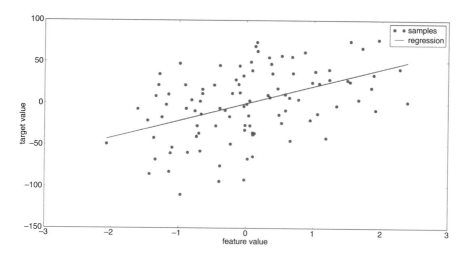

Fig. 3 Sample representation of linear model

3. In the third cell, input training data and output (target) data are initialized.
4. Placeholders for the variables are declared.
5. Compute the equation h = wx + b, compute cost function using the formula.
6. cost = $\dfrac{(h-y)^2}{2m}$ and to minimize the cost, gradient descent optimizer is used.
7. Initialize global variables at once.
8. Run a session (Figs. 4 and 5).

Implementation of linear regression in tensorflow is pretty straightforward. All that is needed is three lines of code. The first line multiplies matrix features to matrix weights. The second line is cost or [8] loss function (least squared error). Finally, one stage of gradient descent optimization is performed by the third line to minimize cost function (Fig. 6).

Many Names of Linear Regression
It can get confusing when you start looking at linear regression [9] because linear regression backs for more than 200 years and still every possible angle of it is being studied. A new different name is assigned to every angle [10]. The regression is an example of a linear model, which presumes a linear correlation among input and output variables. The input variable is denoted as input(x) and output(Y). To measure Y in a more precise manner, a linear arrangement of input variables(x) can be utilized. In general, if there is a single input(x) and single output variable then it is termed as simple linear regression, if the method has a single input variable and has various output variables then it is termed as multiple linear regression.

To improve the performance of linear regression, several techniques are developed and in that the most popular is ordinary square regression also known as least square regression or as least squares linear regression [11].

```
In [1]:  import tensorflow as tf
         import numpy as np
         import matplotlib.pyplot as plt
```

```
In [2]:  training_epoch = 1000
         learning_rate=0.1
         display_step=50
```

```
In [3]:  m=27
         x_train = np.asarray([8,9,7,2,1,3,6,5,4,2,2,3,9,8,5,6,7,3,8,9,2,1,3,6,7,8,9])
         y_train = np.asarray([7,2,3,2,4,5,6,8,7,9,2,5,4,1,7,8,5,9,3,1,3,5,6,7,7,9,4])
```

```
In [4]:  w =tf.Variable(np.random.randn(),name="weight")
         b = tf.Variable(np.random.randn(),name="bias")
         X=tf.placeholder(dtype=tf.float32)
         Y=tf.placeholder(dtype=tf.float32)
```

```
In [5]:  h = tf.add(tf.multiply(w,X),b)
         cost_function = tf.reduce_sum(tf.pow((h-Y),2))/(2*m)
         grad = tf.train.GradientDescentOptimizer(learning_rate).minimize(cost_function)
```

```
In [6]:  init=tf.global_variables_initializer()
```

Fig. 4 Linear regression Part 1

Learning Model

In linear regression, coefficients used in the equations are estimated. Basically, it takes four techniques to prepare and improve the model.

Simple Linear Regression

A simple linear regression has only one input variable. So, only one coefficient needs to be estimated in simple linear regression [12]. This means that standard deviations, correlation, and co-variance are calculated. All of them should be calculated to traverse and analyze the statistical properties of the data.

Ordinary Least Squares

The sum of squared errors can be minimized using the ordinary least squares. It calculates the distance between each value and the regression line is calculated. These distances are squared and all these are added together and finally the sum of squared errors is calculated [13].

```
In [7]: with tf.Session() as sess:
            sess.run(init)
            for epoch in range(training_epoch):
                for i,j in zip(x_train,y_train):
                    sess.run(grad,feed_dict={X:i,Y:j})
                if((epoch+1)%display_step == 0):
                    cost=sess.run(cost_function, feed_dict={X: x_train, Y: y_train})
                    print("Epoch:", '%05d' % (epoch + 1), "cost=", "{:.11f}".format(cost), "W=", sess.run(w), "b=", sess.run(b))
            W=sess.run(w)
            B=sess.run(b)
            cost=sess.run(cost_function,feed_dict={X:x_train,Y:y_train})
        plt.scatter(x_train,y_train)
        plt.plot(x_train,W*x_train+B,color='red')
        plt.show()
```

```
Epoch: 00050 cost= 3.11869812012 W= 0.09532769 b= 4.613852
Epoch: 00100 cost= 2.98443222046 W= -0.046926536 b= 5.6597247
Epoch: 00150 cost= 2.99336028099 W= -0.098279506 b= 6.0372787
Epoch: 00200 cost= 3.00406289101 W= -0.11681731 b= 6.1735716
Epoch: 00250 cost= 3.00890088081 W= -0.1235096 b= 6.222774
Epoch: 00300 cost= 3.01077508926 W= -0.12592548 b= 6.240536
Epoch: 00350 cost= 3.01146721840 W= -0.12679751 b= 6.246947
Epoch: 00400 cost= 3.01171970367 W= -0.12711157 b= 6.249256
Epoch: 00450 cost= 3.01181173325 W= -0.12722653 b= 6.250101
Epoch: 00500 cost= 3.01184320450 W= -0.12726481 b= 6.2503834
Epoch: 00550 cost= 3.01185321808 W= -0.12727836 b= 6.250482
Epoch: 00600 cost= 3.01185321808 W= -0.12727836 b= 6.250482
Epoch: 00650 cost= 3.01185321808 W= -0.12727836 b= 6.250482
Epoch: 00700 cost= 3.01185321808 W= -0.12727836 b= 6.250482
Epoch: 00750 cost= 3.01185321808 W= -0.12727836 b= 6.250482
Epoch: 00800 cost= 3.01185321808 W= -0.12727836 b= 6.250482
Epoch: 00850 cost= 3.01185321808 W= -0.12727836 b= 6.250482
Epoch: 00900 cost= 3.01185321808 W= -0.12727836 b= 6.250482
Epoch: 00950 cost= 3.01185321808 W= -0.12727836 b= 6.250482
Epoch: 01000 cost= 3.01185321808 W= -0.12727836 b= 6.250482
```

Fig. 5 Linear regression Part 2

Ordinary Least Squares (OLS) is used to minimize this expression. It uses linear operations of algebra to minimize. Ample amount of memory must be available for its matrix operation. The OLS is compatible when the errors are serially uncorrelated in the linear estimator.

Gradient Descent

To find the global optimal point of a function, gradient descent is used. Gradient descent assigns a random value to the coefficients of the input variable and error is calculated every time this step iterates [14].

To further minimize the errors, the learning rate can be used to determine the parameters/coefficients and these coefficients are updated from time to time so that the goal can be achieved.

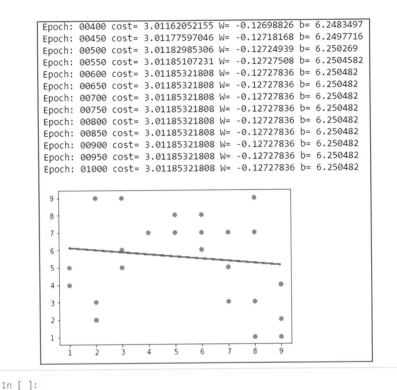

```
Epoch: 00400 cost= 3.01162052155 W= -0.12698826 b= 6.2483497
Epoch: 00450 cost= 3.01177597046 W= -0.12718168 b= 6.2497716
Epoch: 00500 cost= 3.01182985306 W= -0.12724939 b= 6.250269
Epoch: 00550 cost= 3.01185107231 W= -0.12727508 b= 6.2504582
Epoch: 00600 cost= 3.01185321808 W= -0.12727836 b= 6.250482
Epoch: 00650 cost= 3.01185321808 W= -0.12727836 b= 6.250482
Epoch: 00700 cost= 3.01185321808 W= -0.12727836 b= 6.250482
Epoch: 00750 cost= 3.01185321808 W= -0.12727836 b= 6.250482
Epoch: 00800 cost= 3.01185321808 W= -0.12727836 b= 6.250482
Epoch: 00850 cost= 3.01185321808 W= -0.12727836 b= 6.250482
Epoch: 00900 cost= 3.01185321808 W= -0.12727836 b= 6.250482
Epoch: 00950 cost= 3.01185321808 W= -0.12727836 b= 6.250482
Epoch: 01000 cost= 3.01185321808 W= -0.12727836 b= 6.250482
```

In [ ]:

Fig. 6 Linear regression Part 3

Keeping the learning rate too high or too low effects the time of execution and the chance of relevant value of the parameters responsible for minimum error alters. Choosing an appropriate learning rate improves the steps to take on each iteration.

Regularization

Regularization simply adds more information to solve a problem like over-fitting. This step helps to minimize the complexity of the model and also minimizes the miscalculations that take place while training the model using the training data [15].

Following are the examples of regularization in linear regression:

- LASSO regression: It stands for least absolute shrinkage selector operator. It performs L1 regularization [16]. LASSO penalizes the sum of absolute values of the coefficient, which is also called L1 penalty.
- Ridge regression: Here, OLS loss function is modified and the sum of the squared coefficients is minimized [17]. It reduces the model complexity as well as prevents over-fitting.

3 Logistic Regression

Like all regression analyses, the logistic regression is used to compute or predict the probability of the occurrence of a binary or categorical event. The logistic regression algorithm is used for classification. It is used when the target outcome or dependent variable is binary or categorical in nature and is determined by a collection of independent variables. The independent variables could be continuous, ordinal (some given order on a scale), or nominal (named groups). The logistic function helps to threshold the output to either of the two possible binary outcomes in the case of binary logistic regression and to the predetermined number of categories in case of multinomial or ordinal logistic regression.

One example is to predict the possibility of an email being spam (1) or not spam (0). Another example is detecting whether a glioblastoma or a specific type of brain tumor is malignant (1) or benign (0).

Given a scenario, where a need arises to classify an email into spam or non-spam, we can use a linear regression approach to solve the problem. A threshold needs to be established, based on which classification may be performed. If the threshold is chosen as 0.5 with outcomes above the threshold representing malignant tumor and the ground truth of the data point is malignant, with the predicted value of 0.4, the data point will be misclassified as benign, which can lead to serious consequences.

It can be inferred from the above example that linear regression is unsuitable for the above-stated classification problem. Linear regression is unbounded and this summons logistic regression into picture.

Logistic Regression Types

Binary logistic regression: binary logistic regression has only two 2 possible outcomes.

Example: Spam or not spam.

Multinomial logistic regression: Multinomial logistic regression has three or more categorical outcomes.

Example: Predicting which food is preferred more.

Ordinal logistic regression: It can be called as a generalization of binomial logistic regression. Ordinal regression is used if the outcome is ordinal and the proportional odds assumption is met.

For example: A study explores factors that affect the decision of applying to graduate school given in Fig. 7. College juniors are asked to apply for graduate school if they are doubtful, very probable, or very likely to.

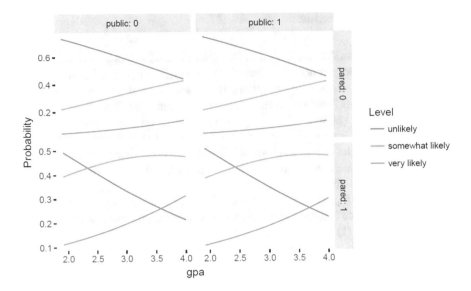

Fig. 7 Logistic regression levels

4 Linear Regression by Importing Datasets

Figure 8 elucidates the linear regression steps for imported data.

1. Import pandas, numpy, tensorflow, random, sklearn.metrics, os.path, sklearn. metrics.
2. Load the datsets using read_csvfrompandas.
3. Initialize TotalFeatures, FeaturesInUse, TotalSampleSize, MiniTestSampleSize.
4. Preprocessing of the data.
5. Initializing the variables.
6. Finding the entropy and defining the optimizer.
7. Training the model.
8. Testing the model.

The dataset contains a total of 376 features where 150 are used.

5 Logistic Regression by Importing Dataset

Figure 10 elucidates the code for logistic regression using seaborn and tensorflow packages.

1. Import numpy, pandas, matplotlib.pyplot, tensorflow.
2. Read the input data (Iris Dataset).
3. Label the data.

```
In [1]: import pandas as pd
        import numpy as np
        from random import sample
        from sklearn.metrics import r2_score
        import os.path
        from sklearn.preprocessing import LabelEncoder
        import tensorflow as tf
```

```
In [7]: test_data=pd.read_csv('../test.csv')
        train_data=pd.read_csv('../train.csv')
```

```
In [8]: TotalFeatures = 376 # we have '376' features
        FeaturesInUse = 150 # number of features used for prediction starting from feature[0]
        TotalSampleSize = 4209 # there are total 4209 training samples.
        MiniTestSampleSize = 409 # lets keep 'MiniTestSampleSize' samples as test data
```

```
In [9]: def PreProcessInputData(test_data, train_data) :
            # first lets process columns, apply labelEncoder to map features to integers
            # thanks to : https://www.kaggle.com/budhiraja/python-pca-regression-baseline-0-5613
            for c in train_data.columns:
                if train_data[c].dtype == 'object':
                    lbl = LabelEncoder()
                    lbl.fit(list(train_data[c].values) + list(test_data[c].values))
                    train_data[c] = lbl.transform(list(train_data[c].values))
                    test_data[c] = lbl.transform(list(test_data[c].values))
                    # you may normalize values here if you want, but again, it didn't helped me much
                    #maxVal = np.amax(pd.concat([test_data[c],train_data[c]], ignore_index=True))
                    #minVal = np.amin(pd.concat([test_data[c],train_data[c]], ignore_index=True))
                    #train_data[c] = (train_data[c] - minVal + 1)/(maxVal + 1)
                    #test_data[c] = (test_data[c] - minVal + 1)/(maxVal + 1)

            y_train_all = train_data['y'].as_matrix();
            y_train_all = np.reshape(y_train_all, (-1, 1));

            x_train_all = train_data.drop(['y','ID'] , axis = 1);
            x_train_all = x_train_all.drop(x_train_all.columns[FeaturesInUse:TotalFeatures], axis = 1); # drop lot few features to keep
            x_train_all = x_train_all.as_matrix();
            x_train_all = np.reshape(x_train_all, (-1, FeaturesInUse));

            # let's randomly pick 'MiniTestSampleSize' samples and keep them as mini test data
            indices = sample(range(len(y_train_all)),MiniTestSampleSize)
            # our mini test data
            y_mini_test = y_train_all[indices]
            x_mini_test = x_train_all[indices]

            # remaining is out training data
            y_train = np.delete(y_train_all, indices, axis=0)
            x_train = np.delete(x_train_all, indices, axis=0)

            # lets processes master test sample data
            x_test_ID = test_data['ID'].as_matrix();
            x_test = test_data.drop('ID', axis = 1);
            x_test = x_test.drop(x_test.columns[FeaturesInUse:TotalFeatures], axis = 1);
            x_test = x_test.as_matrix();
            x_test = np.reshape(x_test, (-1, FeaturesInUse));

            return (x_train, y_train, x_mini_test, y_mini_test, x_test_ID, x_test)
```

```
In [11]: (x_data, y_data, x_mini_test, y_mini_test, test_data_ID, x_test) = PreProcessInputData(test_data, train_data)
```

```
c:\users\manoh\appdata\local\programs\python\python36\lib\site-packages\ipykernel_launcher.py:16: FutureWarning: Method .as_mat
rix will be removed in a future version. Use .values instead.
  app.launch_new_instance()
c:\users\manoh\appdata\local\programs\python\python36\lib\site-packages\ipykernel_launcher.py:21: FutureWarning: Method .as_mat
rix will be removed in a future version. Use .values instead.
c:\users\manoh\appdata\local\programs\python\python36\lib\site-packages\ipykernel_launcher.py:35: FutureWarning: Method .as_mat
rix will be removed in a future version. Use .values instead.
c:\users\manoh\appdata\local\programs\python\python36\lib\site-packages\ipykernel_launcher.py:38: FutureWarning: Method .as_mat
rix will be removed in a future version. Use .values instead.
```

```
In [12]: x = tf.placeholder(tf.float32, [None, FeaturesInUse]);

         # this is our model
         W = tf.Variable(tf.truncated_normal(shape=[FeaturesInUse,1], stddev=0.1))
         b = tf.Variable(tf.constant(0.1, shape=[1]))
         y = tf.matmul(x, W) + b; # predicted values
         y_ = tf.placeholder(tf.float32, [None, 1]); # true y values
```

Fig. 8 Linear regression for imported data

```
In [13]:  cross_entropy = tf.reduce_mean(tf.square(y - y_)); # this is out cost function
          # alternatively, we can define error function to be R^2, but it didn't helped me
          #total_error = tf.reduce_sum(tf.square(tf.subtract(y_, tf.reduce_mean(y_))))
          #unexplained_error = tf.reduce_sum(tf.square(tf.subtract(y_, y)))
          #R_squared = tf.subtract(1.0, tf.div(total_error, unexplained_error))
          #cross_entropy = R_squared

          train_step = tf.train.GradientDescentOptimizerte().minimize(cross_entropy); # lets use adam optimizer

          sess = tf.InteractiveSession(); # start the session
          tf.global_variables_initializer().run();


In [15]:  y_pred = sess.run(y, feed_dict={x: x_test})

          # store to submist.csv
          sub = pd.DataFrame()
          sub['ID'] = test_data_ID
          sub['y'] = y_pred
          sub.to_csv('submit.csv', index=False)



In [14]:  for j in range(2000):
              for i in range( int((TotalSampleSize-MiniTestSampleSize)/200) ): # lets take 200 samples at time
                  batch_x = x_data[i*200:i*200+200]
                  batch_y = y_data[i*200:i*200+200]
                  sess.run(train_step, feed_dict={x: batch_x, y_: batch_y})
              # let's print prediction error for our mini test data
              if 0 == (j%10) :
                  # test error
                  y_pred = sess.run(y, feed_dict={x: x_mini_test})
                  TestError = r2_score(y_mini_test, y_pred)
                  # training error
                  y_pred = sess.run(y, feed_dict={x: x_data})
                  TrainError = r2_score(y_data, y_pred)
                  # print
                  print("for iteration : {:06d}".format(j), " <TestErr> : {:10.4f}".format(TestError), " <TrainError> : {:10.4f}".format(T
              # you might want to break loop here by some means...say you found test error starts increasing
```

```
for iteration : 000590  <TestErr> :    0.3515  <TrainError> :    0.2160
for iteration : 000600  <TestErr> :    0.3604  <TrainError> :    0.2284
for iteration : 000610  <TestErr> :    0.3691  <TrainError> :    0.2402
for iteration : 000620  <TestErr> :    0.3774  <TrainError> :    0.2516
for iteration : 000630  <TestErr> :    0.3854  <TrainError> :    0.2626
for iteration : 000640  <TestErr> :    0.3932  <TrainError> :    0.2732
for iteration : 000650  <TestErr> :    0.4007  <TrainError> :    0.2834
for iteration : 000660  <TestErr> :    0.4079  <TrainError> :    0.2931
for iteration : 000670  <TestErr> :    0.4149  <TrainError> :    0.3026
for iteration : 000680  <TestErr> :    0.4216  <TrainError> :    0.3116
for iteration : 000690  <TestErr> :    0.4281  <TrainError> :    0.3203
for iteration : 000700  <TestErr> :    0.4344  <TrainError> :    0.3287
for iteration : 000710  <TestErr> :    0.4404  <TrainError> :    0.3367
```

Fig. 8 (continued)

4. Visualize the data.
5. Split the training and testing data.
6. Normalize processing.
7. Build the model framework.
8. Train the model.
9. Visualize again.

Predicted Output Dataset

The dataset used is Iris Dataset, and it contains 4 features, namely, SepalLength, SepalWidth, PetalLength, and PetalWidth. Species is the target attribute (Figs. 9 and 10).

Figure 11 represents the train and test accuracy of the dataset.

Fig. 9 Imported dataset for linear regression

	A	B	C
1	ID	y	
2	1	71.58832	
3	2	96.18916	
4	3	80.75554	
5	4	77.81617	
6	5	108.1532	
7	8	90.72318	
8	10	110.3519	
9	11	95.78572	
10	12	116.5672	
11	14	91.26348	
12	15	116.212	
13	16	105.0168	
14	17	93.99651	
15	19	95.60112	
16	20	107.4329	
17	21	94.09531	
18	22	116.9405	
19	23	94.03545	
20	26	96.6071	
21	28	96.04067	
22	29	94.03545	
23	33	94.03545	
24	35	97.30583	
25	41	93.87971	
26	42	93.62134	
27	43	116.0599	
28	45	104.1695	
29	46	99.91942	
30	51	93.74646	
31	53	87.76588	
32	55	113.8544	
33	56	95.6792	
34	57	106.2038	
35	58	96.94417	
36	59	106.1051	
37	63	109.958	

submit

```
In [21]: import numpy as np # linear algebra
         import seaborn as sns
         sns.set(style='whitegrid')
         import pandas as pd # data processing, CSV file I/O (e.g. pd.read_csv)
         import matplotlib.pyplot as plt
         import tensorflow as tf
         %matplotlib inline
```

Importing the dataset

```
In [23]: iris = pd.read_csv('Iris.csv')
         print("Data Shape:", iris.shape)
)        print(iris.head())

         Data Shape: (150, 6)
            Id  SepalLengthCm  SepalWidthCm  PetalLengthCm  PetalWidthCm      Species
         0   1            5.1           3.5            1.4           0.2  Iris-setosa
         1   2            4.9           3.0            1.4           0.2  Iris-setosa
         2   3            4.7           3.2            1.3           0.2  Iris-setosa
         3   4            4.6           3.1            1.5           0.2  Iris-setosa
         4   5            5.0           3.6            1.4           0.2  Iris-setosa
```

```
In [24]: iris = iris[:100]
         iris.shape
```

```
Out[24]: (100, 6)
```

```
In [25]: iris.Species = iris.Species.replace(to_replace=['Iris-setosa', 'Iris-versicolor'], value=[0, 1])
```

```
In [21]: import numpy as np # linear algebra
         import seaborn as sns
         sns.set(style='whitegrid')
         import pandas as pd # data processing, CSV file I/O (e.g. pd.read_csv)
         import matplotlib.pyplot as plt
         import tensorflow as tf
         %matplotlib inline
```

Importing the dataset

```
In [23]: iris = pd.read_csv('Iris.csv')
         print("Data Shape:", iris.shape)
)        print(iris.head())

         Data Shape: (150, 6)
            Id  SepalLengthCm  SepalWidthCm  PetalLengthCm  PetalWidthCm      Species
         0   1            5.1           3.5            1.4           0.2  Iris-setosa
         1   2            4.9           3.0            1.4           0.2  Iris-setosa
         2   3            4.7           3.2            1.3           0.2  Iris-setosa
         3   4            4.6           3.1            1.5           0.2  Iris-setosa
         4   5            5.0           3.6            1.4           0.2  Iris-setosa
```

```
In [24]: iris = iris[:100]
         iris.shape
```

```
Out[24]: (100, 6)
```

```
In [25]: iris.Species = iris.Species.replace(to_replace=['Iris-setosa', 'Iris-versicolor'], value=[0, 1])
```

Fig. 10 Cross entropy loss plot

```
In [31]: def min_max_normalized(data):
             col_max = np.max(data, axis=0)
             col_min = np.min(data, axis=0)
             return np.divide(data - col_min, col_max - col_min)
```

```
In [32]: # Normalized processing, must be placed after the data set segmentation,
         # otherwise the test set will be affected by the training set
         train_X = min_max_normalized(train_X)
         test_X = min_max_normalized(test_X)
```

```
In [33]: # Begin building the model framework
         # Declare the variables that need to be learned and initialization
         # There are 4 features here, A's dimension is (4, 1)
         A = tf.Variable(tf.random_normal(shape=[4, 1]))
         b = tf.Variable(tf.random_normal(shape=[1, 1]))
         init = tf.global_variables_initializer()
         sess = tf.Session()
         sess.run(init)
```

```
In [34]: # Define placeholders
         data = tf.placeholder(dtype=tf.float32, shape=[None, 4])
         target = tf.placeholder(dtype=tf.float32, shape=[None, 1])
```

```
In [36]: mod = tf.matmul(data, A) + b
```

```
In [37]: # Declare loss function
         # Use the sigmoid cross-entropy loss function,
         # first doing a sigmoid on the model result and then using the cross-entropy loss function
         loss = tf.reduce_mean(tf.nn.sigmoid_cross_entropy_with_logits(logits=mod, labels=target))
```

```
In [38]: learning_rate = 0.003
         batch_size = 30
         iter_num = 1500
```

```
In [39]: opt = tf.train.GradientDescentOptimizer(learning_rate)
         # Define the goal
         goal = opt.minimize(loss)
         # Define the accuracy
         # The default threshold is 0.5, rounded off directly
         prediction = tf.round(tf.sigmoid(mod))
         # Bool into float32 type
         correct = tf.cast(tf.equal(prediction, target), dtype=tf.float32)
         # Average
         accuracy = tf.reduce_mean(correct)
         # End of the definition of the model framework
```

```
In [40]: # Start training model
         # Define the variable that stores the result
         loss_trace = []
         train_acc = []
         test_acc = []
```

```
In [41]: # training model
         for epoch in range(iter_num):
             # Generate random batch index
             batch_index = np.random.choice(len(train_X), size=batch_size)
             batch_train_X = train_X[batch_index]
             batch_train_y = np.matrix(train_y[batch_index]).T
             sess.run(goal, feed_dict={data: batch_train_X, target: batch_train_y})
             temp_loss = sess.run(loss, feed_dict={data: batch_train_X, target: batch_train_y})
             # convert into a matrix, and the shape of the placeholder to correspond
             temp_train_acc = sess.run(accuracy, feed_dict={data: train_X, target: np.matrix(train_y).T})
             temp_test_acc = sess.run(accuracy, feed_dict={data: test_X, target: np.matrix(test_y).T})
             # recode the result
             loss_trace.append(temp_loss)
             train_acc.append(temp_train_acc)
             test_acc.append(temp_test_acc)
             # output
             if (epoch + 1) % 300 == 0:
                 print('epoch: {:4d} loss: {:5f} train_acc: {:5f} test_acc: {:5f}'.format(epoch + 1, temp_loss,
                                                                     temp_train_acc, temp_test_acc))

         epoch:  300 loss: 0.535277 train_acc: 0.912500 test_acc: 0.850000
         epoch:  600 loss: 0.488176 train_acc: 0.975000 test_acc: 0.950000
         epoch:  900 loss: 0.461091 train_acc: 0.987500 test_acc: 1.000000
         epoch: 1200 loss: 0.453735 train_acc: 0.987500 test_acc: 1.000000
         epoch: 1500 loss: 0.395630 train_acc: 0.987500 test_acc: 1.000000
```

Fig. 10 (continued)

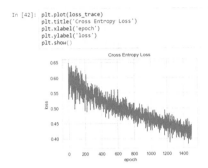

Fig. 10 (continued)

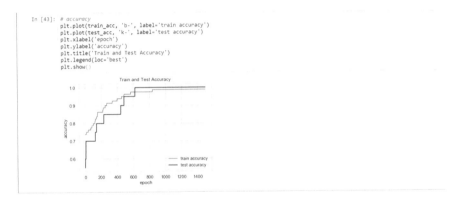

Fig. 11 Train and test accuracy

References

1. C.C. Paige, M. Saunders, LSQR: an algorithm for sparse linear equations and sparse least squares, ACM Transactions on Mathematical Software 8 (1982) 43–71.
2. M. Abramowitz, I.A. Stegun, Handbook of Mathematical Functions, in: National Bureau of Standards, Series, vol. #55, Dover Publications, USA, 1964.
3. R.R. Hocking, Methods and Applications of Linear Models, in: Wiley Series in Probability and Statistics, Wiley-Interscience, New York, 1996.
4. J.R. Magnus, H. Neudecker, Matrix Differential Calculus with Applications in Statistic and Econometrics, revised ed., in: Wiley Series in Probability and Statistics, John Wiley & Sons, Chichester, UK, 1999.
5. G.A.F. Seber, The Linear Hypothesis: A General Theory, in: Griffin's Statistical Monographs and Courses, Charles Griffin and Company Limited, London, 1966.
6. S.F. Ashby, M.J. Holst, T.A. Manteuffel, P.E. Saylor, The role of the inner product in stopping criteria for conjugate gradient iterations, BIT 41 (1) (2001) 26–52.
7. O. Axelsson, I. Kaporin, Error norm estimation and stopping criteria in preconditioned conjugate gradient iterations, Journal of Numerical Linear Algebra with Applications 8 (2001) 265–286
8. Agrawal A, Roy K (2019) Mimicking leaky-integrate-fire spiking neuron using automotion of domain walls for energy-efficient brain-inspired computing. IEEE Trans Magn 55(1):1–7

9. Akinaga H, Shima H (2010) Resistive random access memory (reram) based on metal oxides. Proc IEEE 98(12):2237–2251
10. Amit DJ, Amit DJ (1992) Modeling brain function: the world of attractor neural networks Cambridge University Press, Cambridge
11. Bourzac K (2017) Has intel created a universal memory technology?[news]. IEEE Spectr 54(5):9–10
12. Hands-On Deep Learning for Images with TensorFlow: Build intelligent computer vision applications using TensorFlow and Keras [Authors: Will Ballard]
13. Deep Learning with Applications Using Python: Chatbots and Face, Object, and Speech Recognition with Tensorflow and Keras. [Authors: Navin Kumar Manaswi]
14. Intelligent mobile projects with TensorFlow : build 10+ artificial intelligence apps using TensorFlow Mobile and Lite for iOS, Android, and Raspberry Pi. [Authors: Tang, Jeff]
15. Intelligent Projects Using Python: 9 real-world AI projects leveraging machine learning and deep learning with TensorFlow and Keras. [Authors: SantanuPattanayak]
16. Deep Learning in Python: Master Data Science and Machine Learning with Modern Neural Networks written in Python, Theano, and TensorFlow. [Authors: LazyProgrammer]
17. Deep learning quick reference : useful hacks for training and optimizing deep neural networks with TensorFlow and Keras. [Authors: Bernico, Mike]

Neural Network

Pradeep Kumar Vadla, Adarsha Ruwali, Kolla Bhanu Prakash, M. V. Prasanna Lakshmi, and G. R. Kanagachidambaresan

Here, just assume the word "neuron" simply means a thing that holds a number (specifically a number between 0 and 1) [3].

Following is the program for the neural network to recognize handwritten digits. Modified National Institute of Standards and Technology (MNIST) is the database/dataset consisting of about thousands of sample data in order to train the neural network model [4]. This dataset is made [5] up of images of handwritten digits, 28 × 28 pixels in size as given in Fig. 1. Some examples included in the datasets are below.

1 Inside the Code

One-hot-encoding: One-hot-encoding is used to represent the labels of the images. It uses binary value vectors to represent numerical or categorical values. Our label are the digits (0–9). One of the values is set to 1 to represent the digit at that index of the vector [6] and others are set to 0. For example, the digit 2 is represented by the vector [0, 0, 1, 0, 0, 0, 0, 0, 0, 0].

For representing the actual images, the 28 × 28 pixels are turned (flattened) into a 1D vector that is 784 pixels in [7] size. It is [8] done by X = tf.placeholder("float",[None,n_input]) [the one in the RED].

This Neural Network (NN) model is of 5 layers.

- Input layer: 784 neurons

P. K. Vadla · A. Ruwali · K. B. Prakash · M. V. P. Lakshmi
KL Deemed to be University, Vijayawada, AP, India

G. R. Kanagachidambaresan (✉)
Vel Tech Rangarajan Dr Sagunthala R&D Institute of Science and Technology,
Chennai, Tamil Nadu, India

© Springer Nature Switzerland AG 2021
K. B. Prakash, G. R. Kanagachidambaresan (eds.), *Programming with TensorFlow*, EAI/Springer Innovations in Communication and Computing, https://doi.org/10.1007/978-3-030-57077-4_5

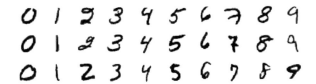

Fig. 1 Handwritten digits 28 × 28 pixels

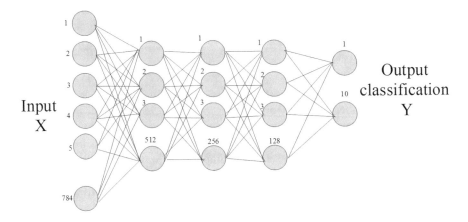

Fig. 2 Visualization of the neural network architecture

- Hidden layer 1: 512 neurons
- Hidden layer 2: 256 neurons
- Hidden layer 3: 128 neurons
- n_output: output layer (0–9digits)

  Figure 2 elucidates the visualization of the neural network architecture:

- Hyper parameters like learning rate, epoch (n_iterations), batch size (to train the large set of data in the model in small chunks), and [9] dropout value (a regularization technique in case if the model overfits the data).
- **Dropout** represents a threshold at which some units are eliminated at random. It will be chosen in our final hidden [10] layer giving each unit a 50% possibility of being eliminated at each training step.

Figure 3 explains the code that is trained in neural network and predicts the hand-written digit image "2."

- **AdamOptimizer** is an optimization algorithm that is used instead of classical stochastic gradient descent procedure update network weights iteratively in training data. It is simply used to minimize cost/loss function (cross-entropy in this case).

The preceding code successfully trains the neural network to classify the MNIST dataset with around 91% accuracy and is tested on an image of digit "2." However, this same problem can achieve about 99% accuracy using a more complex network architecture involving convolutional layers.

NEURAL NETWORK

```python
In [41]: import tensorflow as tf
         from tensorflow.examples.tutorials.mnist import input_data
         import numpy as np
         from PIL import Image
```

```python
In [42]: #store the image data in "mnist" variable
         mnist=input_data.read_data_sets("tmp/data/",one_hot=True)

         #train,validation and test data

         n_train=mnist.train.num_examples
         n_validation=mnist.validation.num_examples
         n_test=mnist.test.num_examples

         Extracting tmp/data/train-images-idx3-ubyte.gz
         Extracting tmp/data/train-labels-idx1-ubyte.gz
         Extracting tmp/data/t10k-images-idx3-ubyte.gz
         Extracting tmp/data/t10k-labels-idx1-ubyte.gz
```

```python
In [15]: #NN layers
         n_input=784 # 28*28 pixels input layer
         n_hidden1=512 #first hidden layer
         n_hidden2=256 #second hidden layer
         n_hidden3=128 #third hidden layer
         n_output=10 #(0-9) digits
```

```python
In [16]: #setting up hyperparameters

         learning_rate=0.0001
         n_iterations=1000
         batch_size=128
         dropout=0.5
```

```python
In [19]: X=tf.placeholder("float",[None,n_input])
         Y=tf.placeholder("float",[None,n_output])
         keep_prob=tf.placeholder(tf.float32)
```

```python
In [20]: weights={'w1': tf.Variable(tf.truncated_normal([n_input,n_hidden1],stddev=0.1)),
                  'w2': tf.Variable(tf.truncated_normal([n_hidden1,n_hidden2],stddev=0.1)),
                  'w3': tf.Variable(tf.truncated_normal([n_hidden2,n_hidden3],stddev=0.1)),
                  'out': tf.Variable(tf.truncated_normal([n_hidden3,n_output],stddev=0.1))
                  }
         biases={'b1':tf.Variable(tf.constant(0.1,shape=[n_hidden1])),
                 'b2':tf.Variable(tf.constant(0.1,shape=[n_hidden2])),
                 'b3':tf.Variable(tf.constant(0.1,shape=[n_hidden3])),
                 'out':tf.Variable(tf.constant(0.1,shape=[n_output])),
                 }
```

```python
In [24]: layer_1=tf.add(tf.matmul(X,weights['w1']),biases['b1'])
         layer_2=tf.add(tf.matmul(layer_1,weights['w2']),biases['b2'])
         layer_3=tf.add(tf.matmul(layer_2,weights['w3']),biases['b3'])
         layer_drop=tf.nn.dropout(layer_3,keep_prob)
         output_layer=tf.matmul(layer_3,weights['out'])+biases['out']
```

Fig. 3 Neural network with handwritten text images

```
In [25]:  #cost/loss function and optimizer to minimize it
          cross_entropy=tf.reduce_mean(tf.nn.softmax_cross_entropy_with_logits(labels=Y, logits=output_layer))
          train_step=tf.train.AdamOptimizer(1e-4).minimize(cross_entropy)

          WARNING:tensorflow:From <ipython-input-25-e0824cf75f17>:2: softmax_cross_entropy_with_logits (from tensorflow.python.ops.nn_op
          s) is deprecated and will be removed in a future version.
          Instructions for updating:

          Future major versions of TensorFlow will allow gradients to flow
          into the labels input on backprop by default.

          See `tf.nn.softmax_cross_entropy_with_logits_v2`.
```

```
In [31]:  #Training and Testing
          correct_pred = tf.equal(tf.argmax(output_layer, 1), tf.argmax(Y, 1))
          accuracy = tf.reduce_mean(tf.cast(correct_pred, tf.float32))

          #initialize session

          init=tf.global_variables_initializer()
          sess=tf.Session()
          sess.run(init)

          #train
          for i in range(n_iterations):
              batch_x,batch_y=mnist.train.next_batch(batch_size)
              sess.run(train_step,feed_dict={X:batch_x, Y:batch_y,keep_prob:dropout})

              #loss and accuracy
              if(i%100==0):
                  minibatch_loss, minibatch_accuracy = sess.run([cross_entropy, accuracy], feed_dict={X: batch_x, Y: batch_y,
                                                                                                       keep_prob:1.0})
                  print("Iteration", str(i), "\t| Loss =", str(minibatch_loss), "\t| Accuracy =", str(minibatch_accuracy))

          test_accuracy=sess.run(accuracy,feed_dict={X:mnist.test.images,Y:mnist.test.labels,keep_prob:1.0})
          print("\nAccuracy on test set: ",test_accuracy)

          Iteration 0    | Loss = 4.3816423    | Accuracy = 0.046875
          Iteration 100  | Loss = 0.4940932    | Accuracy = 0.828125
          Iteration 200  | Loss = 0.35588652   | Accuracy = 0.8984375
          Iteration 300  | Loss = 0.33187914   | Accuracy = 0.890625
          Iteration 400  | Loss = 0.3132461    | Accuracy = 0.9140625
          Iteration 500  | Loss = 0.26009527   | Accuracy = 0.9140625
          Iteration 600  | Loss = 0.20204607   | Accuracy = 0.9375
          Iteration 700  | Loss = 0.25363335   | Accuracy = 0.9140625
          Iteration 800  | Loss = 0.23855169   | Accuracy = 0.9453125
          Iteration 900  | Loss = 0.3274142    | Accuracy = 0.8984375

          Accuracy on test set:  0.9195
```

```
In [40]:  #to test the NN if its actually working or not
          #download any handwritten digits' image of 28 by 28 pixel image
          img = np.invert(Image.open("test_img.png").convert('L')).ravel()

          #predict the given image

          prediction=sess.run(tf.argmax(output_layer,1),feed_dict={X:[img]})
          print("Prediction for test image: ",np.squeeze(prediction))

          Prediction for test image:  2
```

Fig. 3 (continued)

References

1. Python Deep Learning: Exploring deep learning techniques, neural network architectures and GANs with PyTorch, Keras and TensorFlow. [Authors: Ivan Vasilev, Daniel Slater, GianmarioSpacagna, Peter Roelants, Valentino Zocca]
2. Hands-On Transfer Learning with Python Implement Advanced Deep Learning and Neural Network Models Using TensorFlow and Keras [Authors: DipanjanSarkar, Raghav Bali, TamoghnaGhosh]
3. Learning TensorFlow [Authors: Tom Hope, Yehezkel S. Resheff&Itay Lieder]
4. Deep Learning Pipeline: Building A Deep Learning Model With TensorFlow [Authors: Hisham El-Amir, Mahmoud Hamdy]
5. TensorFlow for Machine Intelligence_ A Hands-On Introduction to Learning Algorithms [Authors: Sam Abrahams, DanijarHafner, Erik Erwitt, Ariel Scarpinelli]
6. Python Machine Learning: Machine Learning and Deep Learning with Python, scikit-learn, and TensorFlow [Authors: Sebastian Raschka, VahidMirjalili]

7. Practical Computer Vision Applications Using Deep Learning with CNNs: With Detailed Examples in Python Using TensorFlow and Kivy. [Author: Ahmed Fawzy Gad]
8. Hands-On Machine Learning with Scikit-Learn and TensorFlow: Concepts, Tools, and Techniques to Build Intelligent Systems. [Author: AurélienGéron]
9. Learn TensorFlow 2.0: Implement Machine Learning And Deep Learning Models With thon. [Authors: Pramod Singh, Avinash Manure]
10. Hands-On Deep Learning for Images with TensorFlow: Build intelligent computer vision applications using TensorFlow and Keras [Authors: Will Ballard]

Convolutional Neural Network

Y. V. R. Nagapawan, Kolla Bhanu Prakash, and G. R. Kanagachidambaresan

Although convolutional neural networks (CNNs) have been usually used in image analysis, they can also be used in data analysis and classification problems.

The convolution layer is the main building block of a convolutional neural network. CNNs [3] are neural networks with architectural constraints to reduce computational complexity.

The hidden layers are named convolutional layers. The basis of the CNN are the convolutional layers. Just like any other layers, convolutional layers, the input is received and transformed and output that transformed input to the next layer [4]. With the convolutional layers, this transformation is called convolutional operation. Each convolutional layer has "filters" (simply a matrix with some random values).

1 How Does It Work?

- Basically using CNN, fewer parameters significantly improve the time it takes to learn. It reads the image "chunk-by-chunk" as shown in Fig. 1.
- Influence of nearby pixels is analyzed by a "filter" (also known as "window") that slides over each n × n pixels of input until it slides over every n × n pixels of the image [5]. This sliding is known as "convolving." The amount the filter shifts is named "stride."
- This reduces the number of weights that the neural network must learn compared to a multi-layer perceptron (MLP). Filters are assigned randomly that continuously

Y. V. R. Nagapawan · K. B. Prakash (✉)
KL Deemed to be University, Vijayawada, AP, India

G. R. Kanagachidambaresan
Vel Tech Rangarajan Dr Sagunthala R&D Institute of Science and Technology,
Chennai, Tamil Nadu, India

© Springer Nature Switzerland AG 2021
K. B. Prakash, G. R. Kanagachidambaresan (eds.), *Programming with TensorFlow*, EAI/Springer Innovations in Communication and Computing,
https://doi.org/10.1007/978-3-030-57077-4_6

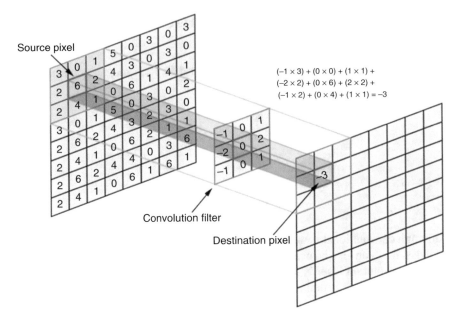

Fig. 1 Stride filter in CNN

update themselves as the network is trained. Figure 2 illustrates the action of edge detection and sharpen filter concepts [6].

- Then, a feature map is generated for each filter that is taken through activation function (ReLu activation function usually) to decide whether a certain feature is present at a given location in the image [7].
- ReLu activation function is used in order to increase the non-linearity in our image. The transition between pixels, the borders, the colors, etc., are the non-linear features.
- Pooling layers are used in order to select the largest values on the feature maps and use these as inputs to other layers. Usually, max pooling is used to find the outliers (an observation point that is different/distant from other observation, Fig. 3).
- In Fig. 3, a cheetah is presented in a different way (normal, rotated, extended). The purpose of max pooling (Fig. 4) is to enable [8] the convolutional neural network to detect the cheetah when presented with the image.
- Now, the pooled feature map is flattened as given in Fig. 5. into a column to insert into an artificial neural network later on [9].

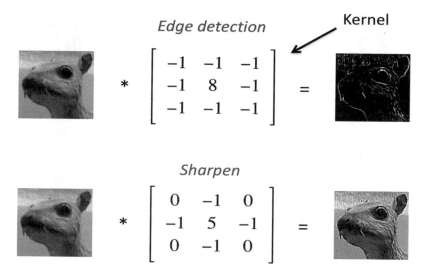

Fig. 2 Edge detection and sharpen filters

Fig. 3 Image of cheetah in (normal, rotated, and extended)

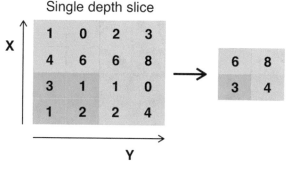

Example of Maxpool with a 2x2 filter and a stride of 2

Fig. 4 Example of 2 x 2 filter with a stride of 2

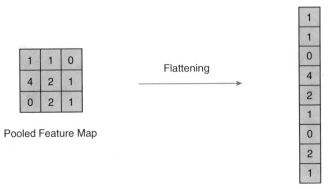

Fig. 5 Flattening the pooled feature map

- **Full Connection:** Now the entire flattened pooled feature map is taken as input to feed into a fully connected neural network [10]. It integrates the features into a wider variety of attributes that [11] improve the ability of convolution network for classifying images.

Here is the example for the digit recognition using convolutional neural network using tensorflow (Fig. 6).

The max training accuracy is about 98.4% as shown in Fig. 7.

CONVOLUTIONAL NEURAL NETWORK

```
In [2]: from __future__ import print_function
        import tensorflow as tf
        from tensorflow.examples.tutorials.mnist import input_data
        import numpy as np
```

```
In [ ]: #store the image data in "mnist" variable
        mnist=input_data.read_data_sets("tmp/data/",one_hot=True)

        # HyperParameters
        learning_rate = 0.001
        training_iters = 10000
        batch_size = 128
        display_step = 10
```

```
In [36]: # Network Parameters
         n_input = 784 #  data input of 28*28 size
         n_classes = 10 # MNIST total classes (0-9 digits)
         dropout = 0.75 # Dropout, probability to keep units
```

```
In [37]: # tf Graph input
         x = tf.placeholder(tf.float32, [None, n_input])
         y = tf.placeholder(tf.float32, [None, n_classes])
         keep_prob = tf.placeholder(tf.float32) #dropout (keep probability)
```

```
In [38]: # filters and ReLu activation function
         def conv2d(x, W, b, strides=1):
             # Conv2D wrapper, with bias and relu activation
             x = tf.nn.conv2d(x, W, strides=[1, strides, strides, 1], padding='SAME')
             x = tf.nn.bias_add(x, b)
             return tf.nn.relu(x)
```

```
In [39]: #MaxPooling
         def maxpool2d(x, k=2):
             # MaxPool2D wrapper
             return tf.nn.max_pool(x, ksize=[1, k, k, 1], strides=[1, k, k, 1],
                                   padding='SAME')
```

```
In [40]: # Create model
         def conv_net(x, weights, biases, dropout):
             # Reshape input picture
             x = tf.reshape(x, shape=[-1, 28, 28, 1])

             # Convolution Layer
             conv1 = conv2d(x, weights['wc1'], biases['bc1'])
             # Max Pooling (down-sampling)
             conv1 = maxpool2d(conv1, k=2)

             # Convolution Layer
             conv2 = conv2d(conv1, weights['wc2'], biases['bc2'])
             # Max Pooling (down-sampling)
             conv2 = maxpool2d(conv2, k=2)

             # Fully connected layer
             # Reshape conv2 output to fit fully connected layer input
             fc1 = tf.reshape(conv2, [-1, weights['wd1'].get_shape().as_list()[0]])
             fc1 = tf.add(tf.matmul(fc1, weights['wd1']), biases['bd1'])
             fc1 = tf.nn.relu(fc1)
             # Apply Dropout
             fc1 = tf.nn.dropout(fc1, dropout)

             # Output, class prediction
             out = tf.add(tf.matmul(fc1, weights['out']), biases['out'])
             return out
```

Fig. 6 Training and accuracy measurement code snippet in Python

```
In [41]:  # Store layers weight & bias
          weights = {
              # 5x5 conv, 1 input, 32 outputs
              'wc1': tf.Variable(tf.random_normal([5, 5, 1, 32])),
              # 5x5 conv, 32 inputs, 64 outputs
              'wc2': tf.Variable(tf.random_normal([5, 5, 32, 64])),
              # fully connected, 7*7*64 inputs, 1024 outputs
              'wd1': tf.Variable(tf.random_normal([7*7*64, 1024])),
              # 1024 inputs, 10 outputs (class prediction)
              'out': tf.Variable(tf.random_normal([1024, n_classes]))
          }

          biases = {
              'bc1': tf.Variable(tf.random_normal([32])),
              'bc2': tf.Variable(tf.random_normal([64])),
              'bd1': tf.Variable(tf.random_normal([1024])),
              'out': tf.Variable(tf.random_normal([n_classes]))
          }

          # Construct model
          pred = conv_net(x, weights, biases, keep_prob)

          # Define loss and optimizer
          cost = tf.reduce_mean(tf.nn.softmax_cross_entropy_with_logits(logits=pred, labels=y))
          optimizer = tf.train.AdamOptimizer(learning_rate=learning_rate).minimize(cost)

          # Evaluate model
          correct_pred = tf.equal(tf.argmax(pred, 1), tf.argmax(y, 1))
          accuracy = tf.reduce_mean(tf.cast(correct_pred, tf.float32))
```

```
In [42]:  # running the graph
          with tf.Session() as sess:
              sess.run(init)
              step = 1
              # Keep training until reach max iterations
              while step * batch_size < training_iters:
                  batch_x, batch_y = mnist.train.next_batch(batch_size)
                  # Run optimization op (backprop)
                  sess.run(optimizer, feed_dict={x: batch_x, y: batch_y,
                                                 keep_prob: dropout})
                  if step % display_step == 0:
                      # Calculate batch loss and accuracy
                      loss, acc = sess.run([cost, accuracy], feed_dict={x: batch_x,
                                                                        y: batch_y,
                                                                        keep_prob: 1.})
                      print("Iter " + str(step*batch_size) + ", Minibatch Loss= " + \
                            "{:.6f}".format(loss) + ", Training Accuracy= " + \
                            "{:.5f}".format(acc))
                  step += 1
              print("Optimization Finished!")

              # Calculate accuracy for 256 mnist test images
              print("Testing Accuracy:", \
                  sess.run(accuracy, feed_dict={x: mnist.test.images[:256],
                                                y: mnist.test.labels[:256],
                                                keep_prob: 1.}))
```

Fig. 6 (continued)

Fig. 7 Training accuracy
value

Iter 44800, Minibatch Loss= 1304.476318, Training Accuracy= 0.95312

Iter 46080, Minibatch Loss= 618.953796, Training Accuracy= 0.94531

Iter 47360, Minibatch Loss= 39.030617, Training Accuracy= 0.98438

References

1. Agrawal A, Roy K (2019) Mimicking leaky-integrate-fire spiking neuron using automotion of domain walls for energy-efficient brain-inspired computing. IEEE Trans Magn 55(1):1–7
2. Akinaga H, Shima H (2010) Resistive random access memory (reram) based on metal oxides. Proc IEEE 98(12):2237–2251
3. Amit DJ, Amit DJ (1992) Modeling brain function: the world of attractor neural networks Cambridge University Press, Cambridge
4. Bourzac K (2017) Has intel created a universal memory technology?[news]. IEEE Spectr 54(5):9–10
5. Deep Learning Pipeline: Building A Deep Learning Model With TensorFlow [Authors: Hisham El-Amir, Mahmoud Hamdy]
6. Goodfellow I, Bengio Y, Courville A, Bengio Y (2016) Deep learning, vol 1. MIT Press, Cambridge
7. Jeong H, Shi L (2018) Memristor devices for neural networks. J Phys D: Appl Phys 52(2):023003
8. Practical Computer Vision Applications Using Deep Learning with CNNs: With Detailed Examples in Python Using TensorFlow and Kivy. [Author: Ahmed Fawzy Gad]
9. Python Deep Learning: Exploring deep learning techniques, neural network architectures and GANs with PyTorch, Keras and TensorFlow. [Authors: Ivan Vasilev, Daniel Slater, Gianmario Spacagna, Peter Roelants, Valentino Zocca]
10. Python Machine Learning: Machine Learning and Deep Learning with Python, scikit-learn, and TensorFlow [Authors: Sebastian Raschka, Vahid Mirjalili]
11. TensorFlow for Machine Intelligence_ A Hands-On Introduction to Learning Algorithms [Authors: Sam Abrahams, Danijar Hafner, Erik Erwitt, Ariel Scarpinelli]

Recurrent Neural Network

G. R. Kanagachidambaresan, Adarsha Ruwali, Debrup Banerjee,
and Kolla Bhanu Prakash

Whenever there is sequence of data like text, speech, video those are connected each after the other [3]. Software like Siri of Apple and Google translate use recurrent neural networks (RNNs). Figure 1 gives the difference in architecture [4] of recurrent/feedback based on feed-forward neural network (NN) architecture. (Sequential data: ordered data that are equally spaced in time.)

1 How They Work?

- In RNN, information cycles through a loop. It takes both the current input and also what it has learned from the inputs [5] it received previously unlike the feed-forward NN.
- Feed-forward NN assigns a weight matrix to its inputs and then produces the output. RNN applies weights to the [6] current as well as the earlier input and adjusts their weight for both gradient descent and backpropagation through time.
- Feed-forward NN maps one input to one output whereas recurrent NN can map many to one (e.g., classifying voice), one to many, and many to many (e.g., translation).

G. R. Kanagachidambaresan (✉)
Vel Tech Rangarajan Dr Sagunthala R&D Institute of Science and Technology,
Chennai, Tamil Nadu, India

A. Ruwali · D. Banerjee · K. B. Prakash
KL Deemed to be University, Vijayawada, AP, India

© Springer Nature Switzerland AG 2021
K. B. Prakash, G. R. Kanagachidambaresan (eds.), *Programming with TensorFlow*, EAI/Springer Innovations in Communication and Computing,
https://doi.org/10.1007/978-3-030-57077-4_7

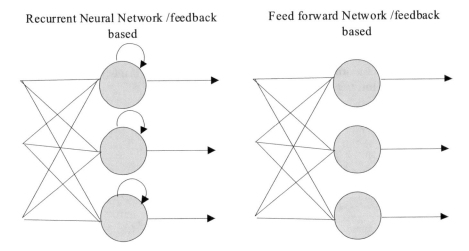

Fig. 1 Recurrent neural network and feed-forwarded neural network

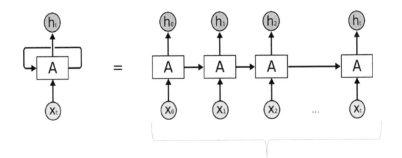

Fig. 2 Unrolled recurrent neural network

1.1 Backpropagation Through Time (BPTT)

- It is simply doing a backpropagation on an unrolled recurrent NN. Unrolling is visualization of RNN to help [7] us understand what is happening within the network.
- While implementing RNN in programming framework, backpropagation is automatically taken care of.

  Figure 2 depicts the unrolled RNN.

- As the error of a present timestep [8] depends on the previous timestep, unrolling of RNN is important for its clarity.
- In BPTT, error is backpropagated from the last to first time step calculating error for each timestep allowing updating the weights.
- However, there are two issues of RNN:

- **Exploding gradients:** When the algorithm assigns considerable importance to the weights, without [9] much reason. This problem can be solved by truncating and squashing the gradients.
- **Vanishing gradients:** When the model stops learning and the values of a gradient is [10] very small. This problem can be solved using long short-term memory (LSTM).

Here, Figs. 3 and 4 program where recurrent NN is implemented using tensor flow.

RECURRENT NEURAL NETWORK

```
In [10]: import tensorflow as tf
         from tensorflow.examples.tutorials.mnist import input_data
         from tensorflow.python.ops import rnn, rnn_cell
         mnist = input_data.read_data_sets("/tmp/data/", one_hot = True)

         Extracting /tmp/data/train-images-idx3-ubyte.gz
         Extracting /tmp/data/train-labels-idx1-ubyte.gz
         Extracting /tmp/data/t10k-images-idx3-ubyte.gz
         Extracting /tmp/data/t10k-labels-idx1-ubyte.gz
```

```
In [11]: hm_epochs = 10
         n_classes = 10
         batch_size = 128
         chunk_size = 28
         n_chunks = 28
         rnn_size = 128

         x = tf.placeholder('float', [None, n_chunks,chunk_size])
         y = tf.placeholder('float')
```

```
In [12]: def recurrent_neural_network(x):
             layer = {'weights':tf.Variable(tf.random_normal([rnn_size,n_classes])),
                     'biases':tf.Variable(tf.random_normal([n_classes]))}

             x = tf.transpose(x, [1,0,2])
             x = tf.reshape(x, [-1, chunk_size])
             x = tf.split(x, n_chunks, 0)

             lstm_cell = rnn_cell.BasicLSTMCell(rnn_size,state_is_tuple=True)
             outputs, states = rnn.static_rnn(lstm_cell, x, dtype=tf.float32)

             output = tf.matmul(outputs[-1],layer['weights']) + layer['biases']

             return output
```

Fig. 3 Tensorflow implementation of RNN part 1

```
In [15]:  def train_neural_network(x):
              prediction = recurrent_neural_network(x)
              cost = tf.reduce_mean( tf.nn.softmax_cross_entropy_with_logits(logits=prediction,labels=y) )
              optimizer = tf.train.AdamOptimizer().minimize(cost)


              with tf.Session() as sess:
                  sess.run(tf.initialize_all_variables())

                  for epoch in range(hm_epochs):
                      epoch_loss = 0
                      for _ in range(int(mnist.train.num_examples/batch_size)):
                          epoch_x, epoch_y = mnist.train.next_batch(batch_size)
                          epoch_x = epoch_x.reshape((batch_size,n_chunks,chunk_size))

                          _, c = sess.run([optimizer, cost], feed_dict={x: epoch_x, y: epoch_y})
                          epoch_loss += c

                      print('Epoch', epoch, 'completed out of',hm_epochs,'loss:',epoch_loss)

                  correct = tf.equal(tf.argmax(prediction, 1), tf.argmax(y, 1))

                  accuracy = tf.reduce_mean(tf.cast(correct, 'float'))
                  print('Accuracy:',accuracy.eval({x:mnist.test.images.reshape((-1, n_chunks, chunk_size)), y:mnist.test.labels}))

          train_neural_network(x)
```

Fig. 4 Tensorflow implementation of RNN during training Part 2

During Training
Figure 5 shows the accuracy in training and testing phase of RNN using Python.

1.2   In the Code

- Tensorflow, mnist, and the rnn [11] model from tensorflow are imported. Chunk size, number of chunks, and rnn size are defined.
- **Recurrent_neural_network(x)** is the function that defines the RNN where weight size is **rnn_size ×n_classes** and biases is [12] just the number of classes (n_classes).
- Input is passed through the LSTM cell that will recur for the **rnn_size.**
- Every cell has **outputs** and **states** at each recurrence [13]. It is done by the following:

    rnn.rnn(lstm_cell, x, dtype = tf.float32)

```
Epoch 0 completed out of 10 loss: 183.533833001
Epoch 1 completed out of 10 loss: 53.2128913924
Epoch 2 completed out of 10 loss: 36.641087316
Epoch 3 completed out of 10 loss: 28.2334972355
Epoch 4 completed out of 10 loss: 23.5787885857
Epoch 5 completed out of 10 loss: 20.3254865455
Epoch 6 completed out of 10 loss: 17.0910299073
Epoch 7 completed out of 10 loss: 15.3585778594
Epoch 8 completed out of 10 loss: 12.5780420878
Epoch 9 completed out of 10 loss: 12.060161829
Accuracy: 0.9827
```

Fig. 5 Tensorflow implementation RNN accuracy level

- In **train_neural_network(x)** function, the default learning rate for AdamOptimizer will be 0.01.
- Accuracy of the trained data is measured up to 98%.

2 Long Short-Term Memory

- LSTM networks are the extension [14] of recurrent NN.
- Allows recurrent NNs to remember their input over a long period of time because it includes memory information that is [15] close to a computer's memory since the LSTM can read, write, and erase information from its memory.
- LSTM consists of three gates: **input**, **forget**, and **output gate**. Well, the work of these are already defined by [16] their names (let new input in, delete information and output at the current timestep). Figure 6 depicts an LSTM cell.

LSTM ranges from 0 to 1. LSTM maintains the gradient svx sufficiently steep and therefore the training is short, and the accuracy is high.

2.1 LSTM In Keras

The following process is the same for vanilla RNN, LSTM, and gated recurrent unit (GRU) when implemented in the keras model.

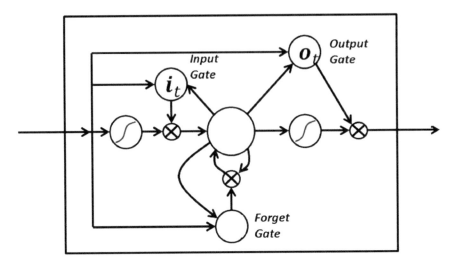

Fig. 6 LSTM gates

```
[54]  from keras.models import Sequential
      from keras.layers import Dense
      from keras.layers import LSTM
      import numpy as np
      import matplotlib.pyplot as plt
      from sklearn.model_selection import train_test_split

[66]  #100 vectors of 5 consecutive digit
      data = [[[(i+j)/100] for j in range(5)] for i in range(100)]
      #converting to numpy arrays
      data = np.array(data, dtype=np.float32)
      #6th consecutive digit of each input sequences
      target = [(i+5)/100 for i in range(100)]
      target = np.array(target, dtype=np.float32)
```

↑ ↓ ⊖ 🗏 ⚙ 🗑

```
#shape
print(data.shape)
print(target.shape)
```

```
(100, 5, 1)
(100,)
```

Fig. 7 Code snippet of LSTM

We want our model to predict the next digit. For instance, if the sequential input is [5, 6, 7, 8, 9], the model should output 10 as given in Fig. 7.

Import the keras-related libraries from keras.models and keras.layers.

- Split training and testing data where 20% of the total size will be used as test data (Fig. 8).
- Add 2 layers of LSTM (Fig. 9):

```
[57]  x_train,x_test,y_train,y_test=train_test_split(data,target,test_size=0.2,random_state=4)
```

Fig. 8 Split training data set

```
[58]  model = Sequential()
      #LSTM layers
      model.add(LSTM(1, batch_input_shape=(None,None,1),return_sequences=True))
      model.add(LSTM((1),return_sequences=False))
```

Fig. 9 Code explanation on LSTM

```
[59]  model.compile(loss='mse', optimizer='adam',metrics=['accuracy'])
```

```
[60]  model.summary()
```

```
⊏→  Model: "sequential_9"
```

Layer (type)	Output Shape	Param #
lstm_14 (LSTM)	(None, None, 1)	12
lstm_15 (LSTM)	(None, 1)	12

```
Total params: 24
Trainable params: 24
Non-trainable params: 0
```

Fig. 10 Code explanation Part 2 LSTM

- *return_sequences = True* would return output after every node
- *return_sequences = False* would return output after last node

- Compile the model with loss function "mean square error(mse)" and [17] optimizer as "adam." *model.summary()* helps to see the output shape and parameters (Fig. 10).
- Fit the model to train with the following parameters as given in Fig. 11.
- Predict the model using *model.predict()* with the test data (x_test) (Fig. 12).
- The plots of predicted digit and the target digit in the test data is shown below (Fig. 13). Some have high difference with the [18] target digit while some are equal.

Similar to what was given above, there are different applications of RNN invariants (RNN, LSTM, GRU) like music generation, stock prediction, and other sequential data with variable input shape and multiple features.

```
[61] model.fit(x_train,y_train, nb_epoch=1000,validation_data=(x_test,y_test))
```

```
Epoch 973/1000
80/80 [==============================] - 0s 370us/step - loss: 0.0629 - ε
Epoch 974/1000
80/80 [==============================] - 0s 449us/step - loss: 0.0625 - ε
Epoch 975/1000
80/80 [==============================] - 0s 469us/step - loss: 0.0621 - ε
Epoch 976/1000
80/80 [==============================] - 0s 457us/step - loss: 0.0617 - ε
Epoch 977/1000
80/80 [==============================] - 0s 451us/step - loss: 0.0613 - ε
Epoch 978/1000
80/80 [==============================] - 0s 441us/step - loss: 0.0608 - ε
Epoch 979/1000
80/80 [==============================] - 0s 440us/step - loss: 0.0604 - ε
Epoch 980/1000
80/80 [------------------------------] - 0s 416us/step - loss: 0.0600    -
```

Fig. 11 Code explanation Part 3 LSTM

```
[62] predict = model.predict(x_test)
```

Fig. 12 Code explanation on mode.predict

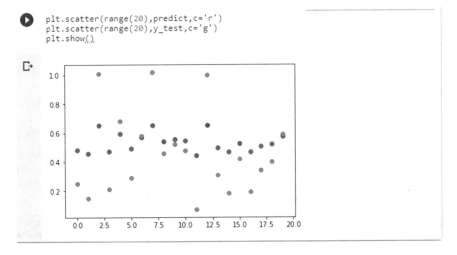

```
plt.scatter(range(20),predict,c='r')
plt.scatter(range(20),y_test,c='g')
plt.show()
```

Fig. 13 Predicted and tested target digit

References

1. Boca Raton Mhaskar H, Liao Q, Poggio T (2016) Learning functions: when is deep better than shallow.arXiv:1603.00988
2. Deep Learning for Computer Vision: Expert techniques to train advanced neural networks using TensorFlow and Keras. [Authors: RajalingappaaShanmugamani]
3. Deep Learning with TensorFlow: Explore neural networks with Python [Authors: Giancarlo Zaccone, Md. RezaulKarim, Ahmed Menshawy]
4. Hands-on unsupervised learning with Python : implement machine learning and deep learning models using Scikit-Learn, TensorFlow, and more [Authors: Bonaccorso, Giuseppe]
5. Li Y, Wang Z, Midya R, Xia Q, Yang JJ (2018) Review of memristor devices in neuromorphic computing: materials sciences and device challenges. J Phys D: ApplPhys 51(50):503002
6. Liao Q, Poggio T (2016) Bridging the gaps between residual learning, recurrent neural networks and visual cortex. arXiv:1604.03640
7. Lippmann R (1987) An introduction to computing with neural nets. IEEE ASSP Mag 4(2):4–22
8. Ma J, Tang J (2017) A review for dynamics in neuron and neuronal network. Nonlinear Dyn 89(3):1569–1578
9. Maan AK, Jayadevi DA, James AP (2017) A survey of memristive threshold logic circuits. IEEE Trans Neural Netw Learn Syst 28(8):1734–1746
10. Mastering TensorFlow 1.x: Advanced machine learning and deep learning concepts using TensorFlow 1.x and Keras. [Author: Armando Fandango]
11. McCulloch WS, Pitts W (1943) A logical calculus of the ideas immanent in nervous activity. Bull Math Biophys 5(4):115–133
12. Medsker L, Jain LC (1999) Recurrent neural networks: design and applications. CRC Press.
13. Practical Computer Vision Applications Using Deep Learning with CNNs: With Detailed Examples in Python Using TensorFlow and Kivy. [Author: Ahmed Fawzy Gad]
14. Practical Deep Learning for Cloud, Mobile, and Edge: Real-World AI & Computer-Vision Projects Using Python, Keras&TensorFlow [Authors: AnirudhKoul, Siddha Ganju, MeherKasam]
15. Python Deep Learning: Exploring deep learning techniques, neural network architectures and GANs with PyTorch, Keras and TensorFlow. [Authors: Ivan Vasilev, Daniel Slater, GianmarioSpacagna, Peter Roelants, Valentino Zocca]
16. Python Machine Learning: Machine Learning and Deep Learning with Python, scikit-learn, and TensorFlow [Authors: Sebastian Raschka, VahidMirjalili]
17. TensorFlow 1.x Deep Learning Cookbook: Over 90 unique recipes to solve artificial-intelligence driven problems with Python. [Authors: Antonio Gulli, AmitaKapoor]
18. TensorFlow for Machine Intelligence_ A Hands-On Introduction to Learning Algorithms [Authors: Sam Abrahams, DanijarHafner, Erik Erwitt, Ariel Scarpinelli]

Application of Machine Learning and Deep Learning

Enireddy Vamsidhar, G. R. Kanagachidambaresan, and Kolla Bhanu Prakash

1   Automobile Industry

Automobile is no exception for the digital transformation. The growing trend in the business and diversification of customer needs has led to innovations in the automobile sector. Machine learning (ML) and deep learning (DL)–based innovations are transforming the automobile industry [1–3]. ML is providing the prediction of automobile needs and usage, quality control, recommendation services, and optimized supply chain management whereas the DL is providing services like automatic lane detection, autonomous driving, and predictive maintenance of the vehicle and nearby service stations (Fig. 1).

2   Climate Change

As the earth keeps warming, the effects of climate change are detrimentally increasing. There were 772 storm and catastrophe incidents in 2016, triple the number that existed in 1980. It threatens the operation of civilization, which undoubtedly needs significant preparation in order to deal with possible changing weather conditions. Weather experts have adopted ML and DL approaches to accelerate the knowledge of different elements of the earth system and associated characteristics. These researches may help for the sustenance of human race. The applications of ML and DL are shown in Fig. 2 [4–6].

E. Vamsidhar · G. R. Kanagachidambaresan
Vel Tech Rangarajan Dr Sagunthala R&D Institute of Science and Technology,
Chennai, Tamil Nadu, India

K. B. Prakash (✉)
KL Deemed to be University, Vijayawada, AP, India

© Springer Nature Switzerland AG 2021
K. B. Prakash, G. R. Kanagachidambaresan (eds.), *Programming with TensorFlow*, EAI/Springer Innovations in Communication and Computing,
https://doi.org/10.1007/978-3-030-57077-4_8

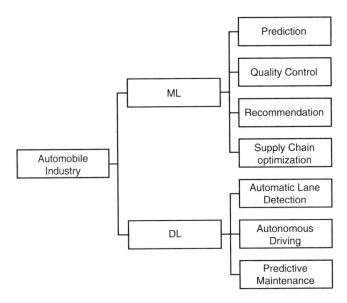

Fig. 1 Applications of ML and DL in automobile industry

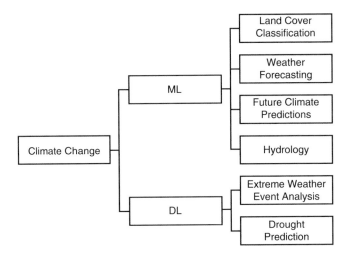

Fig. 2 Applications of ML and DL in climate change

3   Disaster Management

An estimated 1.35 million lives are lost due to disasters over the past 20 years and with post-disaster recovery, approximately 300 U.S. dollars are spent each year. Many catastrophes have an effect on environment, buildings, infrastructure, atmosphere, and local residents [7, 8]. Governments and organizations fail to organize successful strategies for disaster recovery activities. ML and DL are now able to

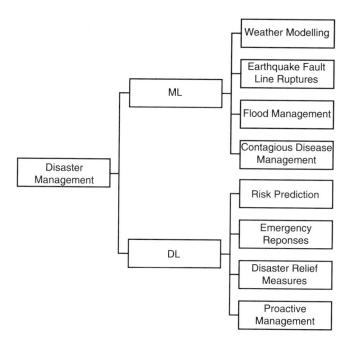

Fig. 3 Applications of ML and DL in disaster management

show their ability in new fields such as disaster management (DM) with their vali-
dated algorithms in detection, estimation, clustering, outer analysis, etc., with real-
time support data from satellites, drones, weather info, etc., as well as history of
disasters. The ML and DL applications of DM are shown in Fig. 3 [9, 10]. It is clear
that the ML and DL techniques will rescue people from awful disasters and also
propose adequate recovery measures to prevent panic [11].

4   Education

ML in education is a method of customized learning that could be used to provide
an individualized educational experience for every student. The students are moti-
vated by their own learning, they must follow the rate they like, and make their own
decisions on what to study according to their curricula. The student assessment,
performance evaluation, grading, and career prediction are not undauntable tasks
with ML and DL algorithms, and feedback on curriculum, teacher, etc., are possible
with less effort provided the data is gathered as per the needs [12–15]. It will take a
while for the old school educators to use ML. But in no time, everyone will realize
that ML will revolutionize the education field and the entire nation [16]. DL has
exciting applications in the world of education [17]. Various applications of ML and
DL are shown in Fig. 4.

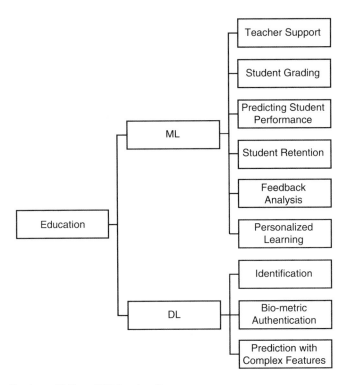

Fig. 4 Applications of ML and DL in education

5 Energy

Carbon sector is the result of a number of sectors. Which comprises coal fuel, electric power, nuclear power, and clean energy industries, along with firewood-based conventional energy industries. Non-renewable fossil and nuclear power are key polluting sources and therefore accountable for global warming. Carbon resource development and use is very important to the global economy. Every economic operation requires energy capital, whether manufacturing products, supplying transportation, operating computers, and other machinery. With ML, the demand prediction, dynamic pricing of the energy as per the demand, recommendation services to the customers, optimal control of generation and distribution, and also extended to disaggregation of energy, that is, separation of profiles of individual receivers from the energy profile signal to better consumption behavior, improve energy efficiency [18, 19]. DL methods are efficiently used in power forecasting [20], preventive diagnostics [21], risk detection [22, 23], etc. Various application of ML and DL in the energy sector are shown in Fig. 5.

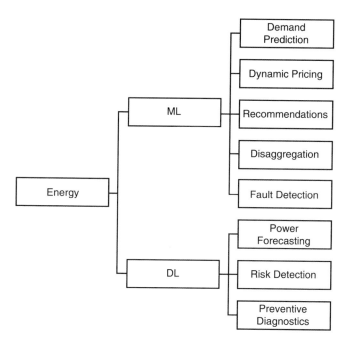

Fig. 5 Applications of ML and DL in energy

6 Entertainment and Media

The prospects for entertainment and media (E&M) to successfully utilize ML are more than amazing. ML and DL are used widely in E&M industry. As rich content is available in internet, the recommender systems gained the ability to recommend the content relevant to the readers, listeners viewers and also the content personalization, target advertisement by companies to the potential customers, personal virtual assistants are intelligent enough to understand the user instructions through voice and optimized video search archives has changed the entertainment and media industry [24–26] (Fig. 6).

The following are a few top research and applications done by different companies for E&M industries in the world.

6.1 AlphaGo

Chess and Go are very popular board games, which resemble in some extent: both are played in turns by two players, and no random factor is involved.

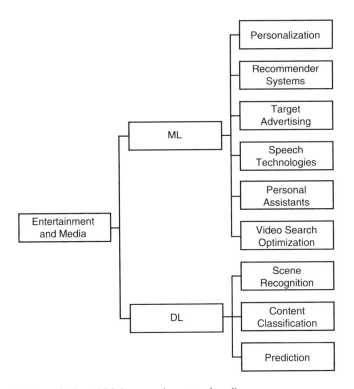

Fig. 6 Applications of ML and DL in entertainment and media

AlphaGo is the artificial intelligence–based (precisely deep reinforcement learning) game playing developed by Google's DeepMind that competes the human grandmasters in the game of GO.

Two different components are used by AlphaGo: convolutional networks that guide the tree search procedure and a tree search procedure. Conceptually, the convolutional networks are somewhat similar to Deep Blue's evaluation function, except they are *not designed* but *learned*. The tree search procedure can be considered as a brute-force approach, while the convolutional networks give the gameplay a level of intuition. Then DeepMind (the London lab behind AlphaGo) released AlphaZero, which defeated the previous 100–0 edition. This edition never focused on a compilation of individual expert gestures and learned solely from the self-play of adversaries.

In total, three Convolutional Neural Network (CNNs) are trained, of two different kinds: two policy networks and one value network in AlphaGo.

6.2 Voice Generation

Google released WaveNet and Baidu released Deep Speech, both of which are DL networks that automatically generate speech.

Until now, text2voice systems have not been fully autonomous in the way they have created new voices; they have been trained to do so manually. The systems that are created nowadays are learning to mimic human voices on their own and improve with time. When you let an audience try to differentiate them from a real human speaking, it is much simpler to do so. While we are not yet there in terms of automated voice generation, DL brings us a step closer to giving machines the ability to speak as human beings do.

6.3 *Music Generation*

A DL network can also be trained to produce music compositions using the same techniques used for voice recognition. Below is an example from Francesco Marchesani who taught the machine how to compose music like the classical composer Chopin. After the computer learns the patterns and statistics unique to Chopin's music, it is creates a whole new piece! For the AI generated music, see https://youtu.be/j60J1cGINX4

6.4 *Restoring Sounds in Video*

A DL network has been trained in a work by Owens et al. on videos where people hit and scratch objects with a drumstick. The scientists muted the video after several learnings and asked the machine to replicate the sound it wants to hear and the results are remarkable: https://youtu.be/0FW99AQmMc8

Similarly, it can also read the lips. LipNet can lip read, accomplished by Oxford and Google's DeepMind scientists. It was 93% accurate in reading people's lips where an average lip reader has the accuracy of 52%. Video: https://youtu.be/fa5QGremQf8

6.5 *Automatically Writing Wikipedia*

Long short-term memory (LSTM) is the DL architecture used here. It is very accurate on textual input. In a blog post called "The Unreasonable Effectiveness of Recurrent Neural Networks" by Andrej Karpathy, a DL model reads math papers, Shakespeare, computer code, and Wikipedia. At the end, the computer wrote just like the Wikipedia articles and also wrote like Shakespeare. Also, the machine could write even computer code and fake math papers. This is a code-writing software for computer programs. The text written by the computer does not make sense all the time, but it is reasonable to expect it to get there.

Figure 7 displays text like Shakespeare's but it was written by a deep network that was fed Shakespeare's writings.

PANDARUS:
Alas, I think he shall be come approached and the day
When little srain would be attain'd into being never fed,
And who is but a chain and subjects of his death,
I should not sleep.

Second Senator:
They are away this miseries, produced upon my soul,
Breaking and strongly should be buried, when I perish
The earth and thoughts of many states.

DUKE VINCENTIO:
Well, your wit is in the care of side and that.

Second Lord:
They would be ruled after this chamber, and
my fair nues begun out of the fact, to be conveyed,
Whose noble souls I'll have the heart of the wars.

Clown:
Come, sir, I will make did behold your worship.

VIOLA:
I'll drink it.

Fig. 7 Deep network fed Shakespeare's words

6.6 Deep-Fake Detection

With the ascendancy of neural network-based learning algorithms, we are now able to take on and defeat problems that sounded completely impossible just a few years ago. One example is creating deep fakes, or in other words we can record a short video and transfer our gestures to target a subject, and this particular technique is so advanced that we do not even need a video of our target, just one still image. A paper by Face Forensics contains a large dataset of original and manipulated video pairs. As this offered a ton of training data for real and forged videos, it became possible to use these to train a deep-fake detector. There is not just on detector algorithm, writing of neural networks have even more variations. Many more AI practitioners created datasets just for deep fake detection [27]. Many politicians have used deep fake for awareness and election campaigns. In 2018, Buzzfeed created a deep fake of Barak Obama with someone else's voice for a public service announcement to increase awareness of deep fakes. In 2020, during the Delhi legislative assembly election campaign, the Bharatiya Janata Party used this technique to distribute a version of an English-language campaign advertisement by their leader.

6.7 Multi-Agent Systems

OpenAI builds a hide-and-seek game for their agents to play with some rules. The agents can move by setting a force on themselves in the x and y directions as well as rotate along the z-axis. The agents can see objects in their line of sight and within a frontal cone. The agents can sense distance to objects, walls, and other agents around them using a LIDAR-like sensor. The main principle behind multi-agents are that they should coordinate, cooperate, and negotiate with each other, much as people do. The goal of this project is to pit two AI teams against each other. The agents can grab and move objects in the environment. The agents can lock objects in place. Only the team that locked an object can unlock it. In the first million rounds, every agents moves around the environment aimlessly without proper strategy and semi-random movements, the seekers are favored, and hence win the majority of the games. Then over time, the hiders learned to lock out the seekers by blocking the doors off with the boxes and started winning consistently. The environment was deliberately designed in the way that hiders can only succeed through collaboration. But after ten million rounds, the seekers learned to move blocks to climb the boxes, and thus seekers started winning [28].

6.8 Image Synthesis

Since many years, the neural network–based technique is mainly used for image classification which means that they were able to recognize objects like animals, traffic, birds, etc., but with the incredible pace of ML, researchers can now have a selection of techniques for not only classification but also synthesizing them. One way of being able to control the output is to use a technique that is capable of image translation, for example, apples to oranges or horses to zebras. It was called CycleGAN because it introduced a cycle consistency loss function which meant that if we convert a summer image to a winter image and then back to a summer image they both should be the same or at least very similar. This can be implemented on images to various applications like upscaling images and to make a beautiful time-lapse video with very smooth translations. Not just changing climate, this implementation can also be used to generate landscapes or terrain images that are used in graphic renders and games [29].

6.9 Graphic Generator

NVIDIA researchers, led by Ting-Chun Wang, have created a new DL method that creates photorealistic images from high-level marks, thus developing a simulated world that enables the consumer to interactively change a scene [30].

7 Finance

The value of ML in finance is becoming increasingly apparent, but the real long-term value will probably only become apparent in the coming years. It is used widely in algorithmic trading, efficient portfolio management, content creation, under writing of loan or insurance, financial risk prediction, sentiment analysis in the stock market, and detecting financial frauds [31–33]. DL improves the accuracy of forecasting [34] in trading and document analysis in the financial institution [31] (Fig. 8).

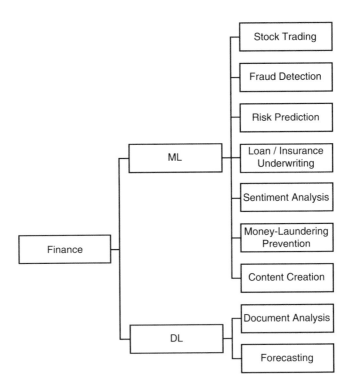

Fig. 8 Applications of ML and DL in energy

References

1. hmian@anaconda.com, "4 Machine Learning Use Cases in the Automotive Sector", https://www.anaconda.com/blog/4-machine-learning-use-cases-automotive, (Last Accessed on 29.04.2020)
2. A. Luckow, M. Cook, N. Ashcraft, E. Weill, E. Djerekarov and B. Vorster, "Deep learning in the automotive industry: Applications and tools," 2016 IEEE International Conference on Big Data (Big Data), Washington, DC, 2016, pp. 3759-3768.
3. Akshay Srinivasa, "4 ways Artificial Intelligence (and Deep Learning) is transforming automotive industry," https://www.pathpartnertech.com/4-ways-artificial-intelligence-and-deep-learning-is-transforming-automotive-industry/ (Last Accessed on 29.04.2020).
4. Chris Huntingford, Elizabeth S Jeffers, Michael B Bonsall, Hannah M Christensen, Thomas Lees and Hui Yang, "Machine learning and artificial intelligence to aid climate change research and preparedness", Environmental Research Letters, 14(12), Nov 2019.
5. Irina Demianchuk, Oxagile, "How Machine Learning and AI Can Help in the Fight Against Climate Change", https://www.technologynetworks.com/informatics/articles/how-machine-learning-and-ai-can-help-in-the-fight-against-climate-change-327269 (Last Accessed on 29.04.2020).
6. RENEE CHO, "Artificial Intelligence—A Game Changer for Climate Change and the Environment", https://blogs.ei.columbia.edu/2018/06/05/artificial-intelligence-climate-environment/, (Last Accessed on 29.04.2020).
7. Alidoost, Fatemeh & Arefi, Hossein. (2018). Application of Deep Learning for Emergency Response and Disaster Management.
8. World Bank, "Machine Learning for Disaster Risk Management", https://reliefweb.int/report/world/machine-learning-disaster-risk-management-guidance-note-how-machine-learning-can-be, (Last Accessed on 29.04.2020).
9. Niki LaGrone, "Ethical Machine Learning for Disaster Relief: Rage for Machine Learning", https://www.azavea.com/blog/2019/12/10/ethical-machine-learning-for-disaster-relief-rage-for-machine-learning/, (Last Accessed on 29.04.2020).
10. Jeff Catlin, "Artificial Intelligence for Disaster Relief: A Primer", https://www.lexalytics.com/lexablog/artificial-intelligence-disaster-relief, (Last Accessed on 29.04.2020).
11. NAVEEN JOSHI. "Machine Learning can save us from disasters!", https://www.allerin.com/blog/machine-learning-can-save-us-from-disasters, (Last Accessed on 29.04.2020).
12. Asthana, Pallavi & Hazela, Bramah. (2020). Applications of Machine Learning in Improving Learning Environment. https://doi.org/10.1007/978-981-13-8759-3_16.
13. Danijel Kučak, Vedran Juričić, Goran Đambić, "Machine Learning In Education - A Survey of Current Research Trends", 29th DAAAM International Symposium on Intelligent Manufacturing and Automation, pp. 406-410, DOI: https://doi.org/10.2507/29th.daaam.proceedings.059
14. Matthew Lynch, "8 Ways Machine Learning Will Improve Education", https://www.thetechedvocate.org/8-ways-machine-learning-will-improve-education, (2018), (Last Accessed on 29.04.2020).
15. Kotsiantis, S.B. Use of machine learning techniques for educational proposes: a decision support system for forecasting students' grades. Artif Intell Rev 37, 331–344 (2012). https://doi.org/10.1007/s10462-011-9234-x
16. How is Machine Learning enhancing the Future of Education?, https://data-flair.training/blogs/machine-learning-in-education/, (Last Accessed on 29.04.2020).
17. ODSC, "Deep Learning is Not Always the Best Solution in Education", https://medium.com/@ODSC/deep-learning-is-not-always-the-best-solution-in-education-3aad239446cf, (Last Accessed on 29.04.2020).
18. Arthur Haponik, "Machine Learning in Energy industry", https://addepto.com/machine-learning-energy-industry/, (Last Accessed on 29.04.2020).

19. Adam Green, "Machine Learning in Energy - A guide for the energy professional", https://towardsdatascience.com/machine-learning-in-energy-c729c1af55c8, (Last Accessed on 29.04.2020).
20. A. Gensler, J. Henze, B. Sick and N. Raabe, "Deep Learning for solar power forecasting — An approach using Auto Encoder and LSTM Neural Networks," 2016 IEEE International Conference on Systems, Man, and Cybernetics (SMC), Budapest, 2016, pp. 002858-002865.
21. Deep Learning on Medium, "6 Applications of Machine Learning in Oil and Gas", https://mc.ai/6-applications-of-machine-learning-in-oil-and-gas/, (Last Accessed on 29.04.2020).
22. F. Chen and M. R. Jahanshahi, "NB-CNN: Deep Learning-Based Crack Detection Using Convolutional Neural Network and Naïve Bayes Data Fusion," in IEEE Transactions on Industrial Electronics, vol. 65, no. 5, pp. 4392-4400, May 2018.
23. S. N. Ahsan and S. A. Hassan, "Machine learning based fault prediction system for the primary heat transport system of CANDU type pressurized heavy water reactor," 2013 International Conference on Open Source Systems and Technologies, Lahore, 2013, pp. 68-74.
24. Top AI and Machine Learning Trends in Media and Entertainment, https://towardsdatascience.com/top-ai-and-machine-learning-trends-in-media-and-entertainment-823f7efea928, (Last Accessed on 29.04.2020).
25. D. Cheng et al., "Scene recognition based on extreme learning machine for digital video archive management," 2015 IEEE International Conference on Robotics and Biomimetics (ROBIO), Zhuhai, 2015, pp. 1619-1624.
26. Stephano Zanetti and Abdennour El Rhalibi. 2004. Machine learning techniques for FPS in Q3. In Proceedings of the 2004 ACM SIGCHI International Conference on Advances in computer entertainment technology (ACE '04). Association for Computing Machinery, New York, NY, USA, 239–244. DOI: https://doi.org/10.1145/1067343.1067374
27. Paratha neekhara, shehzeen Hussain, malhar jere, farinaz koushanfar julian mcauley, "Adversarial Deepfakes: Evaluating Vulnerability of Deepfake Detectors to Adversarial Examples", arXiv:2002.12749v2 [cs.CV] 14 Mar 2020.
28. Browse baker, Ingmar kanitscheider, Todor Markov, Yi Wu, Glenn Powell, Bob McGrew, Ignore Modratch,"Emergent toll use form multi-agent autocurricula", arXiv:1909.07528v2 [cs.LG] 11 Feb 2020.
29. I. Anokhin, P. Solovev, D. Korzhenkov, A. Kharlamov, T. Khakhulin1, A. Silvestrov, S. Nikolenko, V. Lempitsky, G. Sterkin, "High-Resolution Daytime Translation Without Domain Labels", arXiv:2003.08791v2 [cs.CV] 23 Mar 2020.
30. Generating and Editing High-Resolution Synthetic Images with GANs, https://news.developer.nvidia.com/generating-and-editing-high-resolution-synthetic-images-with-gans/, (Last Accessed on 29.04.2020).
31. KC Cheung, "10 Applications of Machine Learning in Finance", https://algorithmxlab.com/blog/applications-machine-learning-finance/, (Last Accessed on 29.04.2020).
32. AVIRAM EISENBERG, "7 Ways Fintechs Use Machine Learning to Outsmart the Competition", https://igniteoutsourcing.com/fintech/machine-learning-in-finance/, (Last Accessed on 29.04.2020).
33. Wei-Yang Lin, Ya-Han Hu, and Chih-Fong Tsai. 2012. Machine Learning in Financial Crisis Prediction: A Survey. Trans. Sys. Man Cyber Part C 42, 4 (July 2012), 421–436, DOI :https://doi.org/ 10.1109/ TSMCC.2011.2170420
34. Min-Yuh Day and Chia-Chou Lee. 2016. Deep learning for financial sentiment analysis on finance news providers. In Proceedings of the 2016 IEEE/ACM International Conference on Advances in Social Networks Analysis and Mining (ASONAM '16). IEEE Press, 1127–1134.

Chatbot

Kolla Bhanu Prakash, A. J. Sravan Kumar, and G. R. Kanagachidambaresan

Significance

Productivity: Chatbots provide the assistance or [2] access to information quickly and efficiently.

Social and relational factors: Conversions of chatbots fuel and strengthen the social experiences. Chatting with bots gives the opportunity [3] for humans to speak without being judged and improves conversational skills.

Customer service and user satisfaction: Digitizing human interactions can be time-efficient and cost-effective so that even smaller firms can afford and maintain good customer service [4].

The following is the code for the chatbot. It is a sequence-to-sequence model trained as [5] a bot. The model is trained on the twitter dataset "twcs.csv". The collected sentences are converted into patterns in the preprocessing [6] part and then fed to our model, which learns relations and representations of the data. The architecture assumes the same prior distributions for input and output words. Therefore, by adopting a new model, it shares [7] embedding layer (pre-trained embedding word) between [8] the encoding and decoding processes. To improve the context sensitivity, the thought vector (i.e., the encoder output) is obtained. To avoid forgetting [9] the context during the response generation, the vector [10] of thought is concatenated with a dense vector encoding the incomplete answer generated up to the current stage.

Code implementation of the chatbot is given in Figs. 1, 2, 3, 4, 5, 6, 7, 8, 9 and 10. Here, 80 iterations are done in total as given in Fig. 8.

K. B. Prakash · A. J. S. Kumar
KL Deemed to be University, Vijayawada, AP, India

G. R. Kanagachidambaresan (✉)
Vel Tech Rangarajan Dr Sagunthala R&D Institute of Science and Technology,
Chennai, Tamil Nadu, India

© Springer Nature Switzerland AG 2021
K. B. Prakash, G. R. Kanagachidambaresan (eds.), *Programming with TensorFlow*, EAI/Springer Innovations in Communication and Computing,
https://doi.org/10.1007/978-3-030-57077-4_9

```python
import re
from time import time

import numpy as np
import pandas as pd
from keras.callbacks import ModelCheckpoint, EarlyStopping, ReduceLROnPlateau
from keras.layers import Dense, Input, LSTM, Embedding, RepeatVector, concatenate, TimeDistributed
from keras.models import Model
from keras.models import load_model
from keras.optimizers import Adam
from keras.utils import np_utils
from nltk.tokenize import casual_tokenize
from sklearn.externals import joblib
from sklearn.feature_extraction.text import CountVectorizer
from sklearn.model_selection import train_test_split


class chatbot:

    def __init__(self):
        self.max_vocab_size = 50000
        self.max_seq_len = 30
        self.embedding_dim = 100
        self.hidden_state_dim = 100
        self.epochs = 80
        self.batch_size = 128
        self.learning_rate = 1e-4
        self.dropout = 0.3
        self.data_path = "twcs.csv"
        self.outpath = ""
        self.version = 'v1'
        self.mode = 'train'
        self.num_train_records = 50000
        self.load_model_from = "s2s_model_v1_.h5"
        self.vocabulary_path = "vocabulary.pkl"
        self.reverse_vocabulary_path = "reverse_vocabulary.pkl"
        self.count_vectorizer_path = "count_vectorizer.pkl"

        self.UNK = 0
        self.PAD = 1
        self.START = 2
```

Fig. 1 Chatbot coding Section 1

```python
def process_data(self, path):
    data = pd.read_csv(path)

    if self.mode == 'train':
        data = pd.read_csv(path)
        data['in_response_to_tweet_id'].fillna(-12345, inplace=True)
        tweets_in = data[data['in_response_to_tweet_id'] == -12345]
        tweets_in_out = tweets_in.merge(data, left_on=['tweet_id'], right_on=['in_response_to_tweet_id'])
        return tweets_in_out[:self.num_train_records]
    elif self.mode == 'inference':
        return data

def replace_anonymized_names(self, data):

    def replace_name(match):
        cname = match.group(2).lower()
        if not cname.isnumeric():
            return match.group(1) + match.group(2)
        return '@__cname__'

    re_pattern = re.compile('(@|^@)([a-zA-Z0-9_]+)')
    if self.mode == 'train':

        in_text = data['text_x'].apply(lambda txt: re_pattern.sub(replace_name, txt))
        out_text = data['text_y'].apply(lambda txt: re_pattern.sub(replace_name, txt))
        return list(in_text.values), list(out_text.values)
    else:
        return list(map(lambda x: re_pattern.sub(replace_name, x), data))

def tokenize_text(self, in_text, out_text):
    count_vectorizer = CountVectorizer(tokenizer=casual_tokenize, max_features=self.max_vocab_size - 3)
    count_vectorizer.fit(in_text + out_text)
    self.analyzer = count_vectorizer.build_analyzer()
    self.vocabulary = {key_: value_ + 3 for key_, value_ in count_vectorizer.vocabulary_.items()}
    self.vocabulary['UNK'] = self.UNK
    self.vocabulary['PAD'] = self.PAD
    self.vocabulary['START'] = self.START
    self.reverse_vocabulary = {value_: key_ for key_, value_ in self.vocabulary.items()}
    joblib.dump(self.vocabulary, self.outpath + 'vocabulary.pkl')
    joblib.dump(self.reverse_vocabulary, self.outpath + 'reverse_vocabulary.pkl')
    joblib.dump(count_vectorizer, self.outpath + 'count_vectorizer.pkl')

def words_to_indices(self, sent):
    word_indices = [self.vocabulary.get(token, self.UNK) for token in self.analyzer(sent)] + [
        self.PAD] * self.max_seq_len
    word_indices = word_indices[:self.max_seq_len]
    return word_indices

def indices_to_words(self, indices):
    return ' '.join(self.reverse_vocabulary[id] for id in indices if id != self.PAD).strip()
```

Fig. 2 Chatbot coding Section 2

```python
def data_transform(self, in_text, out_text):
    X = [self.words_to_indices(s) for s in in_text]
    Y = [self.words_to_indices(s) for s in out_text]
    return np.array(X), np.array(Y)

def train_test_split_(self, X, Y):
    X_train, X_test, y_train, y_test = train_test_split(X, Y, test_size=0.25, random_state=0)
    y_train = y_train[:, :, np.newaxis]
    y_test = y_test[:, :, np.newaxis]
    return X_train, X_test, y_train, y_test

def data_creation(self):
    data = self.process_data(self.data_path)
    in_text, out_text = self.replace_anonymized_names(data)
    test_sentences = []
    test_indexes = np.random.randint(1, self.num_train_records, 10)
    for ind in test_indexes:
        sent = in_text[ind]
        test_sentences.append(sent)
    self.tokenize_text(in_text, out_text)
    X, Y = self.data_transform(in_text, out_text)
    X_train, X_test, y_train, y_test = self.train_test_split_(X, Y)
    return X_train, X_test, y_train, y_test, test_sentences

def define_model(self):

    # Embedding Layer
    embedding = Embedding(
        output_dim=self.embedding_dim,
        input_dim=self.max_vocab_size,
        input_length=self.max_seq_len,
        name='embedding',
    )
    # Encoder input
    encoder_input = Input(
        shape=(self.max_seq_len,),
        dtype='int32',
        name='encoder_input',
    )
    embedded_input = embedding(encoder_input)

    encoder_rnn = LSTM(
        self.hidden_state_dim,
        name='encoder',
        dropout=self.dropout
    )
    # Context is repeated to the max sequence length so that the same context
    # can be feed at each step of decoder
    context = RepeatVector(self.max_seq_len)(encoder_rnn(embedded_input))
```

Fig. 3 Chatbot coding Section 3

```
# Decoder
last_word_input = Input(
    shape=(self.max_seq_len,),
    dtype='int32',
    name='last_word_input',
)
embedded_last_word = embedding(last_word_input)
# Combines the context produced by the encoder and the last word uttered as inputs
# to the decoder.

decoder_input = concatenate([embedded_last_word, context], axis=2)

# return_sequences causes LSTM to produce one output per timestep instead of one at the
# end of the intput, which is important for sequence producing models.
decoder_rnn = LSTM(
    self.hidden_state_dim,
    name='decoder',
    return_sequences=True,
    dropout=self.dropout
)

decoder_output = decoder_rnn(decoder_input)

    # TimeDistributed allows the dense layer to be applied to each decoder output per timestep
    next_word_dense = TimeDistributed(
        Dense(int(self.max_vocab_size / 20), activation='relu'),
        name='next_word_dense',
    )(decoder_output)

    next_word = TimeDistributed(
        Dense(self.max_vocab_size, activation='softmax'),
        name='next_word_softmax'
    )(next_word_dense)

    return Model(inputs=[encoder_input, last_word_input], outputs=[next_word])

def create_model(self):
    _model_ = self.define_model()
    adam = Adam(lr=self.learning_rate, clipvalue=5.0)
    _model_.compile(optimizer=adam, loss='sparse_categorical_crossentropy')
    return _model_
```

Fig. 4 Chatbot coding Section 4

```
# Function to append the START index to the response Y
def include_start_token(self, Y):
    print(Y.shape)
    Y = Y.reshape((Y.shape[0], Y.shape[1]))
    Y = np.hstack((self.START * np.ones((Y.shape[0], 1)), Y[:, :-1]))
    # Y = Y[:,:,np.newaxis]
    return Y

def binarize_output_response(self, Y):
    return np.array([np_utils.to_categorical(row, num_classes=self.max_vocab_size)
                    for row in Y])

def respond_to_input(self, model, input_sent):
    input_y = self.include_start_token(self.PAD * np.ones((1, self.max_seq_len)))
    ids = np.array(self.words_to_indices(input_sent)).reshape((1, self.max_seq_len))
    for pos in range(self.max_seq_len - 1):
        pred = model.predict([ids, input_y]).argmax(axis=2)[0]
        # pred = model.predict([ids, input_y])[0]
        input_y[:, pos + 1] = pred[pos]
    return self.indices_to_words(model.predict([ids, input_y]).argmax(axis=2)[0])
```

```
def train_model(self, model, X_train, X_test, y_train, y_test):
    input_y_train = self.include_start_token(y_train)
    print(input_y_train.shape)
    input_y_test = self.include_start_token(y_test)
    print(input_y_test.shape)
    early = EarlyStopping(monitor='val_loss', patience=10, mode='auto')
```

```
checkpoint = ModelCheckpoint(self.outpath + 's2s_model_' + str(self.version) + '_.h5', monitor='val_loss',
                    verbose=1, save_best_only=True, mode='auto')
lr_reduce = ReduceLROnPlateau(monitor='val_loss', factor=0.5, patience=2, verbose=0, mode='auto')
model.fit([X_train, input_y_train], y_train,
        epochs=self.epochs,
        batch_size=self.batch_size,
        validation_data=[[X_test, input_y_test], y_test],
        callbacks=[early, checkpoint, lr_reduce],
        shuffle=True)
return model
```

```
def generate_response(self, model, sentences):
    output_responses = []
    print(sentences)
    for sent in sentences:
        response = self.respond_to_input(model, sent)
        output_responses.append(response)
    out_df = pd.DataFrame()
    out_df['Tweet in'] = sentences
    out_df['Tweet out'] = output_responses
    return out_df
```

Fig. 5 Chatbot coding Section 5

```
def main(self):
    if self.mode == 'train':
        X_train, X_test, y_train, y_test, test_sentences = self.data_creation()
        print(X_train.shape, y_train.shape, X_test.shape, y_test.shape)
        print('Data Creation completed')
        model = self.create_model()
        print("Model creation completed")
        model = self.train_model(model, X_train, X_test, y_train, y_test)
        test_responses = self.generate_response(model, test_sentences)
        print(test_sentences)
        print(test_responses)
        pd.DataFrame(test_responses).to_csv(self.outpath + 'output_response.csv', index=False)

    elif self.mode == 'inference':
        model = load_model(self.load_model_from)
        self.vocabulary = joblib.load(self.vocabulary_path)
        self.reverse_vocabulary = joblib.load(self.reverse_vocabulary_path)
        # nalyzer_file = open(self.analyzer_path,"rb")
        count_vectorizer = joblib.load(self.count_vectorizer_path)
        self.analyzer = count_vectorizer.build_analyzer()
        data = self.process_data(self.data_path)
        col = data.columns.tolist()[0]
        test_sentences = list(data[col].values)
        test_sentences = self.replace_anonymized_names(test_sentences)
        responses = self.generate_response(model, test_sentences)
        print(responses)
        responses.to_csv(self.outpath + 'responses_' + str(self.version) + '_.csv', index=False)


start_time = time()
obj = chatbot()
obj.mode = "train"
obj.main()
end_time = time()
print("Processing finished, time taken is %s", end_time - start_time)


(37500, 30) (37500, 30, 1) (12500, 30) (12500, 30, 1)
Data Creation completed
Model creation completed
(37500, 30, 1)
(37500, 30)
(12500, 30, 1)
(12500, 30)
Train on 37500 samples, validate on 12500 samples
Epoch 1/80
37500/37500 [==============================] - 182s 5ms/step - loss: 6.3489 - val_loss: 4.8971

Epoch 00001: val_loss improved from inf to 4.89707, saving model to s2s_model_v1_.h5
Epoch 2/80
37500/37500 [==============================] - 179s 5ms/step - loss: 4.6298 - val_loss: 4.4797

Epoch 00002: val_loss improved from 4.89707 to 4.47972, saving model to s2s_model_v1_.h5
Epoch 3/80
37500/37500 [==============================] - 179s 5ms/step - loss: 4.4270 - val_loss: 4.3865
```

Fig. 6 Chatbot coding Section 6

```
Epoch 00003: val_loss improved from 4.47972 to 4.38647, saving model to s2s_model_v1_.h5
Epoch 4/80
37500/37500 [==============================] - 179s 5ms/step - loss: 4.3067 - val_loss: 4.2615

Epoch 00004: val_loss improved from 4.38647 to 4.26150, saving model to s2s_model_v1_.h5
Epoch 5/80
37500/37500 [==============================] - 179s 5ms/step - loss: 4.1799 - val_loss: 4.1384

Epoch 00005: val_loss improved from 4.26150 to 4.13841, saving model to s2s_model_v1_.h5
Epoch 6/80
37500/37500 [==============================] - 179s 5ms/step - loss: 4.0331 - val_loss: 3.9769

Epoch 00006: val_loss improved from 4.13841 to 3.97685, saving model to s2s_model_v1_.h5
Epoch 7/80
37500/37500 [==============================] - 179s 5ms/step - loss: 3.8716 - val_loss: 3.8210

Epoch 00007: val_loss improved from 3.97685 to 3.82105, saving model to s2s_model_v1_.h5
Epoch 8/80
37500/37500 [==============================] - 179s 5ms/step - loss: 3.7219 - val_loss: 3.6875

Epoch 00008: val_loss improved from 3.82105 to 3.68751, saving model to s2s_model_v1_.h5
Epoch 9/80
37500/37500 [==============================] - 178s 5ms/step - loss: 3.5952 - val_loss: 3.5827

Epoch 00009: val_loss improved from 3.68751 to 3.58272, saving model to s2s_model_v1_.h5
Epoch 10/80
37500/37500 [==============================] - 179s 5ms/step - loss: 3.4925 - val_loss: 3.4966

Epoch 00010: val_loss improved from 3.58272 to 3.49657, saving model to s2s_model_v1_.h5
Epoch 11/80
37500/37500 [==============================] - 179s 5ms/step - loss: 3.4034 - val_loss: 3.4258
```

Fig. 7 Chatbot coding Section 7

```
(37500, 30) (37500, 30, 1) (12500, 30) (12500, 30, 1)
Data Creation completed
Model creation completed
(37500, 30, 1)
(37500, 30)
(12500, 30, 1)
(12500, 30)
Train on 37500 samples, validate on 12500 samples
Epoch 1/80
37500/37500 [==============================] - 182s 5ms/step - loss: 6.3489 - val_loss: 4.8971

Epoch 00001: val_loss improved from inf to 4.89707, saving model to s2s_model_v1_.h5
Epoch 2/80
37500/37500 [==============================] - 179s 5ms/step - loss: 4.6298 - val_loss: 4.4797

Epoch 00002: val_loss improved from 4.89707 to 4.47972, saving model to s2s_model_v1_.h5
Epoch 3/80
37500/37500 [==============================] - 179s 5ms/step - loss: 4.4270 - val_loss: 4.3865
```

Fig. 8 Chatbot coding Section 8

```
Epoch 00003: val_loss improved from 4.47972 to 4.38647, saving model to s2s_model_v1_.h5
Epoch 4/80
37500/37500 [==============================] - 179s 5ms/step - loss: 4.3067 - val_loss: 4.2615

Epoch 00004: val_loss improved from 4.38647 to 4.26150, saving model to s2s_model_v1_.h5
Epoch 5/80
37500/37500 [==============================] - 179s 5ms/step - loss: 4.1799 - val_loss: 4.1384

Epoch 00005: val_loss improved from 4.26150 to 4.13841, saving model to s2s_model_v1_.h5
Epoch 6/80
37500/37500 [==============================] - 179s 5ms/step - loss: 4.0331 - val_loss: 3.9769

Epoch 00006: val_loss improved from 4.13841 to 3.97685, saving model to s2s_model_v1_.h5
Epoch 7/80
37500/37500 [==============================] - 179s 5ms/step - loss: 3.8716 - val_loss: 3.8210

Epoch 00007: val_loss improved from 3.97685 to 3.82105, saving model to s2s_model_v1_.h5
Epoch 8/80
37500/37500 [==============================] - 179s 5ms/step - loss: 3.7219 - val_loss: 3.6875

Epoch 00008: val_loss improved from 3.82105 to 3.68751, saving model to s2s_model_v1_.h5
Epoch 9/80
37500/37500 [==============================] - 178s 5ms/step - loss: 3.5952 - val_loss: 3.5827

Epoch 00009: val_loss improved from 3.68751 to 3.58272, saving model to s2s_model_v1_.h5
Epoch 10/80
37500/37500 [==============================] - 179s 5ms/step - loss: 3.4925 - val_loss: 3.4966

Epoch 00010: val_loss improved from 3.58272 to 3.49657, saving model to s2s_model_v1_.h5
Epoch 11/80
37500/37500 [==============================] - 179s 5ms/step - loss: 3.4034 - val_loss: 3.4258

Epoch 00070: val_loss improved from 2.72088 to 2.71803, saving model to s2s_model_v1_.h5
Epoch 71/80
37500/37500 [==============================] - 181s 5ms/step - loss: 2.0399 - val_loss: 2.7164

Epoch 00071: val_loss improved from 2.71803 to 2.71642, saving model to s2s_model_v1_.h5
Epoch 72/80
37500/37500 [==============================] - 180s 5ms/step - loss: 2.0310 - val_loss: 2.7160

Epoch 00072: val_loss improved from 2.71642 to 2.71596, saving model to s2s_model_v1_.h5
Epoch 73/80
37500/37500 [==============================] - 181s 5ms/step - loss: 2.0204 - val_loss: 2.7145

Epoch 00073: val_loss improved from 2.71596 to 2.71448, saving model to s2s_model_v1_.h5
Epoch 74/80
37500/37500 [==============================] - 181s 5ms/step - loss: 2.0119 - val_loss: 2.7119

Epoch 00074: val_loss improved from 2.71448 to 2.71186, saving model to s2s_model_v1_.h5
Epoch 75/80
37500/37500 [==============================] - 181s 5ms/step - loss: 2.0028 - val_loss: 2.7098

Epoch 00075: val_loss improved from 2.71186 to 2.70983, saving model to s2s_model_v1_.h5
Epoch 76/80
37500/37500 [==============================] - 181s 5ms/step - loss: 1.9937 - val_loss: 2.7093
```

Fig. 9 Chatbot coding Section 9

```
Epoch 00076: val_loss improved from 2.70983 to 2.70925, saving model to s2s_model_v1_.h5
Epoch 77/80
37500/37500 [==============================] - 180s 5ms/step - loss: 1.9849 - val_loss: 2.7089

Epoch 00077: val_loss improved from 2.70925 to 2.70885, saving model to s2s_model_v1_.h5
Epoch 78/80
37500/37500 [==============================] - 181s 5ms/step - loss: 1.9768 - val_loss: 2.7084

Epoch 00078: val_loss improved from 2.70885 to 2.70836, saving model to s2s_model_v1_.h5
Epoch 79/80
37500/37500 [==============================] - 181s 5ms/step - loss: 1.9676 - val_loss: 2.7056

Epoch 00080: val_loss improved from 2.70558 to 2.70196, saving model to s2s_model_v1_.h5
['@Kimpton Gyppy Love is a BAD idea for the goldfish! I would not be able to support a business that did this
(1, 30)
(1, 30)
(1, 30)
(1, 30)
(1, 30)
(1, 30)
(1, 30)
(1, 30)
(1, 30)
(1, 30)
['@Kimpton Gyppy Love is a BAD idea for the goldfish! I would not be able to support a business that did this
                                Tweet in                                              Tweet out
0   @Kimpton Gyppy Love is a BAD idea for the gold...   @__cname__ hi , sorry to hear this . please dm...  ·
1   @TwitterSupport I'm being bullied https://t.co...   @__cname__ hi there ! can you dm us your accou...
2   Alright @__cname__ and @__cname__...FIX THIS C...   @__cname__ hi , we are sorry to hear this . pl...
3   @delta flying home from Britain tomorrow and o...   @__cname__ hi there , i'm sorry to hear this ....
4   @AskPlayStation Hey I just brought a code for ...   @__cname__ hi there ! can you dm us your accou...
5   Banks are suck rip offs!! Screw your #MonthlyM...   @__cname__ hi , we are sorry to hear this . pl...
6   Has anyone else looked at travel offers for Bl...   @__cname__ hi , we are sorry to hear this . pl...
7   @__cname__ your international customer servic...    @__cname__ hi , we are sorry to hear this . pl...
8   Hey @SouthwestAir again more preboards than th...   @__cname__ hi , sorry to hear this . please dm...
9         nao esta sendo facil https://t.co/hxNkzKliva  @__cname__ we apologize for the inconvenience ...
Processing finished, time taken is %s 15171.930249929428
```

Fig. 10 Chatbot coding Section 10

References

1. Deep Learning with Applications Using Python: Chatbots and Face, Object, and Speech Recognition with Tensorflow and Keras. [Authors: Navin Kumar Manaswi]
2. Hands-On Machine Learning with Scikit-Learn and TensorFlow: Concepts, Tools, and Techniques to Build Intelligent Systems. [Author: Aurélien Géron]
3. Learn TensorFlow 2.0: Implement Machine Learning And Deep Learning Models With Python. [Authors: Pramod Singh, Avinash Manure]
4. Hands-On Transfer Learning with Python Implement Advanced Deep Learning and Neural Network Models Using TensorFlow and Keras [Authors: Dipanjan Sarkar, Raghav Bali, Tamoghna Ghosh]
5. Hands-On Deep Learning for Images with TensorFlow: Build intelligent computer vision applications using TensorFlow and Keras [Authors: Will Ballard]
6. Deep Learning with TensorFlow: Explore neural networks with Python [Authors: Giancarlo Zaccone, Md. Rezaul Karim, Ahmed Menshawy]
7. TensorFlow 1.x Deep Learning Cookbook: Over 90 unique recipes to solve artificial-intelligence driven problems with Python. [Authors: Antonio Gulli, Amita Kapoor]

8. Hands-on unsupervised learning with Python : implement machine learning and deep learning models using Scikit-Learn, TensorFlow, and more [Authors: Bonaccorso, Giuseppe]
9. Mastering TensorFlow 1.x: Advanced machine learning and deep learning concepts using TensorFlow 1.x and Keras. [Author: Armando Fandango]
10. Practical Deep Learning for Cloud, Mobile, and Edge: Real-World AI & Computer-Vision Projects Using Python, Keras & TensorFlow [Authors: Anirudh Koul, Siddha Ganju, Meher Kasam]

PyTorch

Sagar Imambi, Kolla Bhanu Prakash, and G. R. Kanagachidambaresan

Python is favored for coding and working with deep learning and thus has a wide range of languages and libraries to look over, as given in Fig. 1.

Theano, one of the main deep learning structures, has stopped dynamic improvement. TensorFlow has devoured Keras altogether [2], elevating it to a top of the line Application Program Interface (AP)I. PyTorch is a scientific computing library, supplanted a large portion of the low-level code reused from the Lua-based Torch venture. Initially, PyTorch was created by Hugh Perkins as a Python wrapper.

It included help for ONNX, a seller unbiased [3] model portrayal, trade design, and a deferred execution "diagram mode" runtime called TorchScript. PyTorch is another deep learning library with the abilities of fast performance. Essentially, it is the Facebook answer for combine burn with Python.

1 The Significant Highlights of PyTorch

Simple Interface − PyTorch offers simple to utilize API; subsequently, it is viewed as extremely easy to work with and runs on Python. The [4] code execution in this structure is effortless.

Python use − This library is viewed as Pythonic, which smoothly incorporates with the Python core functions. In this way, it can use every one of the administrations and functionalities offered by Python [5]. PyTorch ensures local help for Python and utilization of its libraries.

S. Imambi · K. B. Prakash (✉)
KL Deemed to be University, Vijayawada, AP, India

G. R. Kanagachidambaresan
Vel Tech Rangarajan Dr Sagunthala R&D Institute of Science and Technology,
Chennai, Tamil Nadu, India

© Springer Nature Switzerland AG 2021
K. B. Prakash, G. R. Kanagachidambaresan (eds.), *Programming with TensorFlow*, EAI/Springer Innovations in Communication and Computing,
https://doi.org/10.1007/978-3-030-57077-4_10

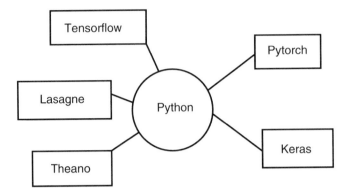

Fig. 1 The Python extension libraries

Dynamic computational charts – PyTorch gives a fantastic stage that offers dynamic computational diagrams. In this way, a client can transform them during runtime. It is a major highlight of PyTorch. This is exceptionally [6] valuable when an engineer has no clue about how a lot of memory is required for making a neural system model. They guarantee the diagram would develop progressively – at each purpose of code execution, the chart is worked along and can be controlled at runtime. So every part of the code executing graph was built and able to manipulate at runtime.

Facebook: It is effectively utilized in the improvement of Facebook for every last bit of its deep learning necessities in the stage. It is actively used in the development of Facebook and its subsidiary companies [7].

FAST: PyTorch is quick and feels local, henceforth guaranteeing simple coding and fast handling.

Compute Unified Device Architecture (CUDA): The help for CUDA guarantees that the code can run on the graphical processor in this way increasing the performance of the network.

2 Why We Prefer PyTorch

- It is simple to debug and comprehend the code.
- Has the same number of sort of layers as Torch (Unpool, Convolution (CONV) 1,2,3D, Long Short Term Memory networks (LSTM), Grus).

- A variety of loss functions are available.
- It can be considered as a numpy augmentation to GPUs.
- It is quicker than other libraries, as chainer and dynet.

3 Requirements for Implementing Deep Learning

Deep learning calculations are intended to vigorously rely upon very good quality machines as opposed to conventional Artificial Intelligence (AI) [8] calculations. Deep learning calculations play out a lot of network duplication tasks that require tremendous equipment support. To execute the PyTorch programs, we require a PC or laptop with a CUDA-competent graphical processing unit (GPU), GPU with 8GB of RAM (we recommend an NVIDIA GTX 1070 or better).

4 PyTorch Basic Components

PyTorch is known for having three levels of abstraction as given below:

- Tensor: N-dimensional array which runs on GPU.
- Variable: Node in computational graph. This stores data and gradient.
- Module: Neural network layer that will store state or learnable weights.

4.1 Tensor

PyTorch has an inside data structure, the tensor, a multi-dimensional group that offers various resemblances with numpy. From [9] that foundation, apparel overviews of features have been created that make it easy to prepare an endeavor for the activity or an assessment concerning another neural framework building organized and arranged.

Tensors give animating of logical assignments and PyTorch has packs for passed on getting ready, expert structures for capable data stacking, and an expansive library of typical significant learning limits.

Creating Tensors
After installing the packages, initialize the empty tensor and assign required values.

Ex:

```
#creates an empty matrix of size 5 × 3
a=torch.empty(5,3)
#Initilizematix with zeros
a=torch r.zeroes(5,3,dtype= torch.long)
#assign values to x
a=torch.tensor(5)
```

Data Type of Elements

PyTorch automatically decides the data type of the elements of the tensor when it is created; the data type applies to all the [10] elements of the tensor. Sometimes that can be overridden to convert it into another data type.

Ex:

```
b = torch.tensor([[3, 8, 9],[4-6]])
print(b.dtype)
# torch.int64
x= torch.tensor([[1,2.5,3],[5.3,5,6]])
print(x.dtype)
 # torch.float32
```

While initializing also we can define data type:

```
b = torch.tensor([[3, 8, 9],[4-6]], dtype=torch.int32)
print(b.dtype)
 # torch.int32
```

Creating Torch tensors from numpy array.

Ex:

```
a= np.ones(4)
b=torch.from_numpy(a)
```

Here *a* is numpy array and initialized with ones. When torch tensor is created using numpy array, they share underlying memory location.

Creating numpy array from torch tensors:

Ex:

```
a= torch.ones(4)
b=a.numpy()
```

4.2 Autograd Module

The most important thing PyTorch offers is to apply auto differentiation. We will see how it works in Fig. 2.

Basic operations are based on the training data set. Next we reply all the values of data to reduce the loss at every stage [11]. Then compute gradients. Gradients are computed by finding the negative slope and calculating the minima of the loss function. Automatic differentiation is a difficult and complicated process and that is easy through the autograd module. This module created the dynamic computational graphs as given in Fig. 3.

Fig. 2 Steps for autograd Module

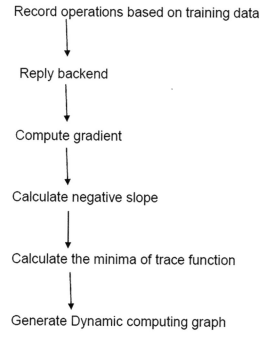

Record operations based on training data

Reply backend

Compute gradient

Calculate negative slope

Calculate the minima of trace function

Generate Dynamic computing graph

```
# creating tensors
x = torch.tensor(1., requires grad=True)
w = torch.tensor(1.5, requires grad=True)
b = torch.tensor(3., requires grad=True)

# creating computational graph.
y = w * x + b    # y = 2 * x + 3

# calculating gradients.
y.backward()

# Print out the gradients.
print(x.grad)    # x.grad = 2
print(w.grad)    # w.grad = 1
print(b.grad)    # b.grad = 1
```

Fig. 3 Dynamic computational graphs

5 Implement the Neural Network Using PyTorch

Training a deep learning algorithm involves the following steps: Building a data pipeline, building network architecture, using [12] loss function to evaluate the architecture, and optimizing the weights of the network architecture using an optimization algorithm.

Preparing a deep learning program includes the accompanying advances like building an information pipeline, building [13] system design, evaluating the engineering utilizing a loss function, and optimizing the weights of the network by an optimizing algorithm as given Fig. 4.

```
# step 1 import package and library
import torch

import torch.nn as nn

# step 2 creating data

x = torch.randn(batch_size, n_in)

y = torch.tensor([[1.0], [0.0], [0.0],
[1.0], [1.0], [1.0], [0.0], [0.0], [1.0], [1.0]])

# step 3 Create a model

model = nn.Sequential(nn.Linear(n_in, n_h),
  nn.ReLU(),
  nn.Linear(n_h, n_out),
  nn.Sigmoid())

criterion = torch.nn.MSELoss()

# step4 Construct the optimizer   layer

optimizer = torch.optim.SGD(model.parameters(), lr = 0.01)

for epoch in range(50):
    u_pred = model(x)

# Step 5: Compute loss loss function

  loss = criterion(u_pred, y)

  print('epoch: ', epoch,' loss: ', loss.item())

  optimizer.zero_grad()

  loss.backward()

# Step 6. Run the autograde

    optimizer.step()
```

Fig. 4 Code snippet for optimizing algorithm

Neural network may be implemented simply by these steps:

Step 1: Import package and libraries.
Step 2: Input data.
Step 3: Construct NN using torch.nn package.
Step 4: Define all layers.
Step 5: Construct loss function.
Step 6: Run autograd.

The output generated of the above program is as given in Fig. 5.

6 Difference Between PyTorch and Tensorflow

Table 1 illustrates the classification of PyTorch and Tensorflow.

```
epoch: 0 loss: 0.2545787990093231

epoch: 1 loss: 0.2545052170753479

epoch: 2 loss: 0.254431813955307

epoch: 3 loss: 0.25435858964920044

epoch: 4 loss: 0.2542854845523834

epoch: 5 loss: 0.25421255826950073

epoch: 6 loss: 0.25413978099823
```

Fig. 5 Output screenshot of the above code

Table 1 PyTorch and Tensorflow classification

S.no.	PyTorch	Tensorflow
1	Dynamic computational graph	Static computational graph
2	Can make use of standard Python flow control	Not able to use
3	Supports Python debugging	Does not support
4	Dynamic inspection of variable and gradients	Not possible
5	Research oriented	Product oriented
6	Developed by Facebook group	Developed by Google group

7 PyTorch for Computer Vision

Computer vision is absolutely one of the fields that [14] has been generally affected by the appearance of profound learning, for an assortment of reasons. The requirement for characterizing or translating the substance of regular pictures existed, enormous datasets became accessible, and convolution layers were created that could be run rapidly on GPUs [15] with remarkable exactness. This joined with the inspiration of the Internet mammoths to comprehend pictures shot by a large number of clients through their cell phones and oversaw on said goliaths' platforms.

7.1 Image Classifier

Image classifier predicts data based [16] on an image set by constructing a neural network. Character/object recognition is generally an image processing technique where image data is imputed and explored by various libraries of Python and PyTorch.

Exploring Data
Standard Python package can be used to load data into numpy array. Then it can be converted into torch tensor. Image data is [17] converted using pillow, opencv. Audio data is interpreted using scipy and librosa and text data is by spacy and cython, etc.

The image prediction using the PyTorch networks is as given in Fig. 6.

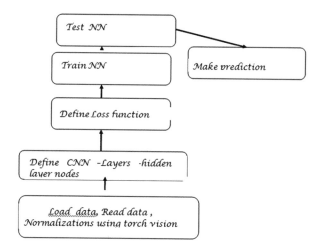

Fig. 6 Steps for image classifier

Data Loading

The first step in deep [18] learning is information loading and handling. PyTorch offers utilities for the identical in torch.utils.data. The crucial training in this module is Dataset and DataLoader. Dataset is built on the pinnacle of tensor data type and is used often for user defined datasets. DataLoader is used if you have a massive dataset and you need to load information from a Dataset in historical past in order that it is equipped and looking ahead to the schooling loop. We can also use torch. nn.DataParallel and torch.distributed if CUDA is available. Figure 7 elucidates the code snippet of data loading.

Defining layers and hidden nodes of network is given in Figs. 8 and 9.

```
imagedir = 'data/images

images = {x: datasets.ImageFolder(os.path.join(imagedir, x))

    for i in ['train', 'val']}

imageloaders = {i: torch.utils.data.DataLoader(images[i], batch  size=4,

                      shuffle=True, num  workers=4)

    for i in ['train', 'val']}

Images  size = {i: len(images [i]) for i in ['train', 'val']}

classlabels = images ['train'].classes


device = torch.device("cuda:0" if torch.cuda.is  available() else "cpu")
```

Fig. 7 Code snippet of data loading

Fig. 8 Layer definition

```
self.linear_layers = Sequential(
    Linear(128 * 14 * 14, 512),
    ReLU(inplace=True),
    Dropout(),
    Linear(512, 256),
    ReLU(inplace=True),
    Dropout(),
    Linear(256,10),
    ReLU(inplace=True),
    Dropout(),
    Linear(10,2)
)
```

```
# Checking whether cuda is available or not
Cudab = torch.cuda.is available()
# Creating  model
model = SimoleNet(num classes=10)
#whenever cuda is available, change the model to the GPU
if cudab:
    model.cuda()
#initilizing optimizer and loss function
opt = Adam(model.parameters(), lr=0.001, weight decay=0.0001)
loss_function = nn.CrossEntropyLoss()
```

Fig. 9 CUDA implementation of algorithm

```
Def validating()
  model.eval()
  test_acc = 0.0
  for x (imgs, cnames) in enumerate(test_loader):
      if cuda_avail:
          imgs = Variable(imgs.cuda())
          cnames = Variable(cnames.cuda())

      outputs = model(imgs)
      _, prediction = torch.max(outputs,data, 1)

      accuracy += torch.sum(prediction == cnames,data)
  # avg accuracy for 10000 images
  accuracy = accuracy / 10000
  return accuracy
```

Fig. 10 Code snippet for accuracy

Next we have to define optimizer and check whether CUDA is available. If it is then use GPU model.

The model was trained after providing information [19] like batch size and number of epochs. Then validating and testing, the model can be done (Fig. 10).

7.2   Image Augmentation in Less Data

We can utilize picture increase for profound learning in any setting – hackathons, industry ventures, etc. We will likewise construct a picture order model utilizing PyTorch to see how picture growth fits into the image.

Deep learning models as a rule require [20] a ton of information for preparing. As a rule, the more the information, the better the exhibition of the model. Be that as it may, obtaining monstrous measures of information accompanies its own difficulties. Not every person has the profound pockets of the enormous firms.

And the issue with an absence of information is [21] that our profound learning model will probably not take in the example or capacity from the information and henceforth it will probably not give a decent presentation on inconspicuous information.

Image augmentation is the way toward producing new pictures for preparing our profound learning model. These new pictures are created utilizing the current preparing pictures and consequently we do not need to gather them physically.

Various Image Augmentation Techniques

Image Rotation

Picture revolution is one of the most ordinarily utilized expansion procedures. It can enable our model to get hearty to the adjustments in the direction of items. Regardless of whether we pivot the picture, the data of the picture continues as before. A vehicle is a vehicle regardless of whether we see it from an alternate point

Shifting/Moving Images

There may be situations when the articles in the picture are not consummately focal adjusted. In these cases, picture move [22] can be utilized to add move invariance to the pictures. By moving the pictures, we can change the situation of the article in the picture and consequently give more assortments to the model. This will in the end lead to a progressively summed up model.

Flipping Images

Flipping is an augmentation of turn. It enables us to flip the picture in the left directly just as up-down bearing. We should perceive how we can execute flipping.

After applying the various operations on the images, the data set is ready for the building model. The process is same as the image classification model as now the data set is having sufficient data.

8 Sequential Data Models

Natural language processing (NLP) provides endless opportunities for artificial intelligence problem solving, making products like Amazon Alexa and Google Translate possible. If you are a developer or data scientist new to NLP and deep learning, this hands-on guide will teach you how to use these approaches with PyTorch, a deep learning application built in Python.

Now, we have seen various feed-forward systems. That is, there is no situation any stretch of imagination keeps up by the system. This is probably not the conduct that we need. Grouping models are vital for NLP: They are models where there is a kind of reliance between your data sources over time. The traditional case of a grouping model is the hidden Markov model for grammatical feature labeling. Another model is the restrictive arbitrary field.

An intermittent neural system is a system that keeps up some sort of state. For instance, its yield could be utilized as a component of the following info, with the goal that data can propagate along as the system disregards the arrangement. On account of an LSTM, for every component in the succession, there is a comparing shrouded state, which on a fundamental level can contain data from subjective focuses prior in the arrangement. We can utilize the concealed state to foresee words in a language model, grammatical feature labels, and a bunch of different things.

8.1 *LSTM in PyTorch*

Note a few things before you get into the example. The LSTM at PyTorch finds all its inputs to be 3D tensors. The semantics of those tensors "axes" are essential. The first axis is the series itself, the second [23] one indexes the mini-batch instances, and the third one indexes the input elements. We have not discussed mini-batching, so let us just ignore that and assume that on the second axis we will always have only 1 dimension. If we want to run the model sequence over the phrase "The girl walks," our input should look as follows (Figs. 11 and 12):

[girl, is, walking]

Ex 2:
An LSTM for part-of-speech tagging.

Problem Definition Part-of-speech labeling is an outstanding assignment in natural language processing. It alludes to the way toward ordering words into their grammatical forms (otherwise called word classes or lexical classifications). This is an administered learning approach as LSTM is an extension of RNN where it shows the inputs of participants.

```
lstm = nn.LSTM(3, 3)   # Input dim is 3, output dim is 3
inputs = [torch.randn(1, 3) for _ in range(5)]   # make a sequence of length 5

# initialize the hidden state.
hidden = (torch.randn(1, 1, 3),
          torch.randn(1, 1, 3))
for i in inputs:
    # Step through the sequence one element at a time.
    # after each step, hidden contains the hidden state.
    out, hidden = lstm(i.view(1, 1, -1), hidden)

# alternatively, we can do the entire sequence all at once.
# the first value returned by LSTM is all of the hidden states throughout
# the sequence. the second is just the most recent hidden state
# (compare the last slice of "out" with "hidden" below, they are the same)
# The reason for this is that:
# "out" will give you access to all hidden states in the sequence
# "hidden" will allow you to continue the sequence and backpropagate,
# by passing it as an argument  to the lstm at a later time
# Add the extra 2nd dimension
inputs = torch.cat(inputs).view(len(inputs), 1, -1)
hidden = (torch.randn(1, 1, 3), torch.randn(1, 1, 3))   # clean out hidden state
out, hidden = lstm(inputs, hidden)
print(out)
print(hidden)
```

Fig. 11 LSTM implementation in PyTorch Part 1

```
tensor([[[-0.0187,  0.1713, -0.2944]],

        [[-0.3521,  0.1026, -0.2971]],

        [[-0.3191,  0.0781, -0.1957]],

        [[-0.1634,  0.0941, -0.1637]],

        [[-0.3368,  0.0959, -0.0538]]], grad_fn=<StackBackward>)
(tensor([[[-0.3368,  0.0959, -0.0538]]], grad_fn=<StackBackward>),
tensor([[[-0.9825,  0.4715, -0.0633]]], grad_fn=<StackBackward>))
```

Fig. 12 LSTM implementation in PyTorch Part 2

However, there is an additional second dimension with size 1.

A unique integral was assigned to each term (and tag). We measure a set of unique words (and tags), then turn them into a list, and index them into a dictionary [24]. The word vocabulary and the tag vocabulary are those dictionaries. We will also add a special padding value for the sequences (more on that later), and another one for unknown vocabulary words (Fig. 13).

```
def prepare_sequence(seq, to_ix):
    idxs = [to_ix[w] for w in seq]
    return torch.tensor(idxs, dtype=torch.long)


training_data = [
    ("The dog ate the apple".split(), ["DET", "NN", "V", "DET", "NN"]),
    ("Everybody read that book".split(), ["NN", "V", "DET", "NN"])
]
word_to_ix = {}
for sent, tags in training_data:
    for word in sent:
        if word not in word_to_ix:
            word_to_ix[word] = len(word_to_ix)
print(word_to_ix)
tag_to_ix = {"DET": 0, "NN": 1, "V": 2}

# These will usually be more like 32 or 64 dimensional.
# We will keep them small, so we can see how the weights change as we train.
EMBEDDING_DIM = 6
HIDDEN_DIM = 6
```

Fig. 13 PyTorch implementation on vocabulary words

```
class LSTMTagger(nn.Module):

    def __init__(self, embedding_dim, hidden_dim, vocab_size, tagset_size):
        super(LSTMTagger, self).__init__()
        self.hidden_dim = hidden_dim

        self.word_embeddings = nn.Embedding(vocab_size, embedding_dim)

        # The LSTM takes word embeddings as inputs, and outputs hidden states
        # with dimensionality hidden_dim.
        self.lstm = nn.LSTM(embedding_dim, hidden_dim)

        # The linear layer that maps from hidden state space to tag space
        self.hidden2tag = nn.Linear(hidden_dim, tagset_size)

    def forward(self, sentence):
        embeds = self.word_embeddings(sentence)
        lstm_out, _ = self.lstm(embeds.view(len(sentence), 1, -1))
        tag_space = self.hidden2tag(lstm_out.view(len(sentence), -1))
        tag_scores = F.log_softmax(tag_space, dim=1)
        return tag_scores
```

Fig. 14 Code implementation of LSTM tagger

In preprocessing the sequence is generated. And words are generated from the sentences using nltk library. The tokens are indexed and assigned the corresponding index as given in Fig. 14.

```
model = LSTMTagger(EMBEDDING_DIM, HIDDEN_DIM, len(word_to_ix), len(tag_to_ix))
loss_function = nn.NLLLoss()
optimizer = optim.SGD(model.parameters(), lr=0.1)

# See what the scores are before training
# Note that element i,j of the output is the score for tag j for word i.
# Here we don't need to train, so the code is wrapped in torch.no_grad()
with torch.no_grad():
    inputs = prepare_sequence(training_data[0][0], word_to_ix)
    tag_scores = model(inputs)
    print(tag_scores)
```

Fig. 15 Code implementation of parameter tuning

```
for epoch in range(300):  # again, normally you would NOT do 300 epochs, it is toy
data
    for sentence, tags in training_data:
        # Step 1. Remember that Pytorch accumulates gradients.
        # We need to clear them out before each instance
        model.zero_grad()

        # Step 2. Get our inputs ready for the network, that is, turn them into
        # Tensors of word indices.
        sentence_in = prepare_sequence(sentence, word_to_ix)
        targets = prepare_sequence(tags, tag_to_ix)

        # Step 3. Run our forward pass.
        tag_scores = model(sentence_in)

        # Step 4. Compute the loss, gradients, and update the parameters by
        #  calling optimizer.step()
        loss = loss_function(tag_scores, targets)
        loss.backward()
        optimizer.step()
```

Fig. 16 Epoch initialization and model fit

The model was built using the parameter tuned (Fig. 15).
The process is repeated until we get the best fit model (Figs. 16 and 17).

9 Summary

- Deep network models robotically learn to associate inputs and favored outputs
  from examples.
- Libraries like PyTorch can help you construct and educate neural network fash-
  ions efficiently.

```
tensor([[-1.1389, -1.2024, -0.9693],
        [-1.1065, -1.2200, -0.9834],
        [-1.1286, -1.2093, -0.9726],
        [-1.1190, -1.1960, -0.9916],
        [-1.0137, -1.2642, -1.0366]])
tensor([[-0.0462, -4.0106, -3.6096],
        [-4.8205, -0.0286, -3.9045],
        [-3.7876, -4.1355, -0.0394],
        [-0.0185, -4.7874, -4.6013],
        [-5.7881, -0.0186, -4.1778]])
```

Fig. 17 Output matrix value

- PyTorch minimizes cognitive overhead, while focusing on flexibility and speed. It additionally defaults to immediate execution for operations.
- TorchScript is a pre-compiled deferred execution mode that can be invoked from Cpp.
- PyTorch gives some of software libraries to facilitate deep studying projects.
- PyTorch is used in a variety of deep learning applications like object detection, image analysis, and sequence modeling.

References

1. Deep Learning for Computer Vision: Expert techniques to train advanced neural networks using TensorFlow and Keras. [Authors: RajalingappaaShanmugamani]
2. Deep Learning in Python: Master Data Science and Machine Learning with Modern Neural Networks written in Python, Theano, and TensorFlow. [Authors: LazyProgrammer]
3. Deep learning quick reference: useful hacks for training and optimizing deep neural networks with TensorFlow and Keras. [Authors: Bernico, Mike]
4. Deep Learning with TensorFlow: Explore neural networks with Python [Authors: Giancarlo Zaccone, Md. RezaulKarim, Ahmed Menshawy]
5. Erdmann M, Glombitza J, Walz D. A deep learning-based reconstruction of cosmic ray-induced air showers. AstropartPhys 2018;97:46–53. doi:https://doi.org/10.1016/j.astropartphys.2017.10.006, URL http://www.sciencedirect.com/science/article/pii/S0927650517302219.
6. Feng Q, Lin TTY. The analysis of VERITAS muon images using convolutional neural networks, in: Proceedings of the International Astronomical Union, vol. 12, 2016.
7. Goodfellow I, Bengio Y, Courville A, Bengio Y (2016) Deep learning, vol 1. MIT Press, Cambridge
8. Grandison T, Sloman M (2000) A survey of trust in internet applications. IEEE CommunSurv Tutor 3(4):2–16

9. Guo X, Ipek E, Soyata T (2010) Resistive computation: avoiding the power wall with low-leakage, STT-MRAM based computing. In: ACM SIGARCH computer architecture news, vol 38. ACM, pp 371–382
10. Hands-on unsupervised learning with Python: implement machine learning and deep learning models using Scikit-Learn, TensorFlow, and more [Authors: Bonaccorso, Giuseppe]
11. Holch TL, Shilon I, Büchele M, Fischer T, Funk S, Groeger N, Jankowsky D, Lohse T, Schwanke U, Wagner P. Probing convolutional neural networks forevent reconstruction in γ -ray astronomy with Cherenkov telescopes, in:PoS ICRC2017, The Fluorescence detector Array of Single-pixel Telescopes:Contributions to the 35th International Cosmic Ray Conference (ICRC2017), p. 795, https://doi.org/10.22323/1.301.0795, arXiv:1711.06298.
12. Huennefeld M. Deep learning in physics exemplified by the reconstruction of muon-neutrino events in IceCube, in: PoS ICRC2017, The Fluorescence detector Array of Single-pixel Telescopes: Contributions to the 35th International Cosmic Ray Conference (ICRC 2017), p. 1057, https://doi.org/10.22323/1.301.1057.
13. Hurst S (1969) An introduction to threshold logic: a survey of present theory and practice. Radio Electron Eng 37(6):339–351
14. Intelligent Projects Using Python: 9 real-world AI projects leveraging machine learning and deep learning with TensorFlow and Keras. [Authors: SantanuPattanayak]
15. Internet of Things for Industry 4.0, EAI, Springer, Editors, G. R. Kanagachidambaresan, R. Anand, E. Balasubramanian and V. Mahima, Springer.
16. Jeong H, Shi L (2018) Memristor devices for neural networks. J Phys D: ApplPhys 52(2):023003
17. Krestinskaya O, Dolzhikova I, James AP (2018) Hierarchical temporal memory using memristor networks: a survey. IEEE Trans Emerg Top ComputIntell 2(5):380–395. doi:https://doi.org/10.1109/TETCI.2018.2838124
18. LeCun Y, Bengio Y, Hinton G. Deep learning. Nature 2015;521(7553):436–44. doi:https://doi.org/10.1038/nature14539.
19. Mangano S, Delgado C, Bernardos M, Lallena M, Vzquez JJR. Extracting gamma-ray information from images with convolutional neural networkmethods on simulated cherenkov telescope array data. In: ANNPR 2018. LNAI, vol. 11081, 2018, p. 243–54. doi:https://doi.org/10.1007/978-3-319-99978-4, arXiv:1810.00592.
20. Mastering TensorFlow 1.x: Advanced machine learning and deep learning concepts using TensorFlow 1.x and Keras. [Author: Armando Fandango]
21. Practical Deep Learning for Cloud, Mobile, and Edge: Real-World AI & Computer-Vision Projects Using Python, Keras&TensorFlow [Authors: AnirudhKoul, Siddha Ganju, MeherKasam]
22. Python Deep Learning: Exploring deep learning techniques, neural network architectures and GANs with PyTorch, Keras and TensorFlow. [Authors: Ivan Vasilev, Daniel Slater, GianmarioSpacagna, Peter Roelants, Valentino Zocca]
23. Shilon I, Kraus M, Büchele M, Egberts K, Fischer T, HolchTL, Lohse T, Schwanke U, Steppa C, Funk S. Application of deep learning methods to analysis of imaging atmospheric cherenkov telescopes data. AstropartPhys 2019; 105: 44–53. doi:https://doi.org/10.1016/j.astropartphys.2018.10.003,   URL   http://www.sciencedirect.com/science/article/pii/S0927650518301178
24. TensorFlow 1.x Deep Learning Cookbook: Over 90 unique recipes to solve artificial-intelligence driven problems with Python. [Authors: Antonio Gulli, AmitaKapoor]

Pattern Recognition and Machine Learning

Bharadwaj, Kolla Bhanu Prakash, and G. R. Kanagachidambaresan

1   Kernel Support Vector Machine

For linearly separable data points and different classes, we can perform simple support vector machine (SVM), but for the data which are non-linear simple (straight line) SVM cannot be suited. The most important point to be noted in the SVM is that it can solve the non-linearly separable problems also. To solve the non-linear problems effectively, two new techniques called soft margin and kernel tricks are introduced.

Soft Margin: A line will separate both the classes and it can also tolerate one or a small number of misclassified dots.

Kernel Trick: It is used to locate a decision boundary in the case of non-linear models.

Kernel SVM is used for non-linearly separable data as it projects non-linearly separable data lower dimensions to linearly separable data in higher dimensions.

Types of Kernels
The most commonly used kernels in the SVM classifier are as follows:

1. Linear kernel
2. Radial basis function (RBF) kernel
3. Polynomial kernel

Bharadwaj · K. B. Prakash (✉)
KL Deemed to be University, Vijayawada, AP, India

G. R. Kanagachidambaresan
Vel Tech Rangarajan Dr Sagunthala R&D Institute of Science and Technology,
Chennai, Tamil Nadu, India

© Springer Nature Switzerland AG 2021
K. B. Prakash, G. R. Kanagachidambaresan (eds.), *Programming with TensorFlow*, EAI/Springer Innovations in Communication and Computing,
https://doi.org/10.1007/978-3-030-57077-4_11

SVM needs to find the optimal line to properly identify any class. RBF and polynomial kernels are the most widely used.

The Gaussian RBF [3] is the most famous and simple RBF kernel. The influence of new features is controlled by gamma (γ). If this value is high then decision on the boundary will be influenced by these features.

Polynomial To draw a non-linear decision boundary, a polynomial kernel can play a vital role in drawing a good solution with the use of polynomial features.

How Does it Work

1. Create pattern matrix.
2. Choose a right kernel function.
3. Choose the kernel function parameter and regularization parameter "C" value.
4. Obtain α by executing training algorithm.
5. Using learned weights α and support vectors unseen data is classified.

Advantages in Using Kernel SVM

1. It works pretty well only in the cases where a clear margin of separation is found between classes.
2. It works well for high dimensional spaces.
3. Risk of over fitting is less.
4. Performs well for good kernel function.

Disadvantages

1. It is not an easy task to find the right kernel.
2. As the dataset size increases the training time also increases.
3. Understanding and interpreting the final model, variable weight is very difficult.
4. SVM hyper parameters are cost (C) and gamma, not easy to tune and hard to visualize.

SVM Applications

1. Protein structure prediction
2. Intrusion detection
3. Handwritten recognition
4. Detecting steganography in digital images
5. Breast cancer diagnosis

1.1 Linear Kernel

Figure 1 elucidates the linear kernel code snippet (Fig. 2).

```
import numpy as np
import matplotlib.pyplot as plt
from sklearn import model_selection
from sklearn import svm, datasets
from sklearn.metrics import classification_report
from sklearn.metrics import confusion_matrix
iris = datasets.load_iris()
X = iris.data[:2, :3]
y = iris.target
svc = svm.SVC(kernel ='linear', C = 1).fit(X, y)
x_min, x_max = X[:, 0].min(), X[:, 0].max() + 1
y_min, y_max = X[:, 1].min(), X[:, 1].max() + 1
h = (x_max / x_min)/100
xx, yy = np.meshgrid(np.arange(x_min, x_max, h), np.arange(y_min, y_max, h))
plt. subplot(1, 1, 1)
Z = svc.predict(np.c_[xx.ravel(), yy.ravel()])
Z = Z.reshape(xx.shape)
plt. contour f(xx, yy, Z, cmap = plt.cm.Paired, alpha = 0.8)
plt.scatter (X[:, 0], X[:, 1], c = y, cmap = plt.cm.Paired)
plt.xlabel('Sepal length')
plt.ylabel('Sepal width')
plt.xlim(xx.min(), xx.max())
plt.title('SVC with linear kernel')
predicted = svc.predict(X)
matrix = confusion_matrix(y,predicted)
print(matrix)
report = classification_report(y,predicted)
print(report)
result=svc.score(X, y)
print("accuracy:%.3f%%"%(result*100.0))
plt.show()
```

Fig. 1 SVC-based linear kernel implementation

Fig. 2 SVC linear kernel   Text (0.5, 1.0, 'SVC with linear kernel')

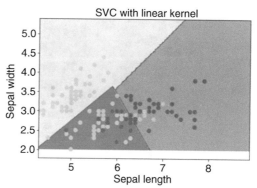

1.2 *RBF Kernel*

Figure 3 elucidates the RBF kernel code using Python.

Figures 4 and 5 elucidate the output distribution [4] graph and accuracy attained through the algorithm.

1.3 *Polynomial Kernel*

Figure 6 illustrates the polynomial kernel implementation [5] code snippet using Python. Figure 7 shows the SVC distribution in the second space.

```python
import numpy as np
import matplotlib.pyplot as plt
from matplotlib.colors import Normalize
from sklearn.svm import SVC
from sklearn.preprocessing import StandardScaler
from sklearn.datasets import load_iris
from sklearn.model_selection import StratifiedShuffleSplit
from sklearn.model_selection import GridSearchCV
class MidpointNormalize(Normalize):
    def __init__(self, vmin=None, vmax=None, midpoint=None, clip=False):
        self.midpoint = midpoint
        Normalize.__init__(self, vmin, vmax, clip)

    def __call__(self, value, clip=None):
        x, y = [self.vmin, self.midpoint, self.vmax], [0, 0.5, 1]
        return np.ma.masked_array(np.interp(value, x, y))

iris = load_iris()
X = iris.data
y = iris. target
 X_2d = X[:, :2]
X_2d = X_2d[y > 0]
y_2d = y[y > 0]
y_2d -= 1
scaler = StandardScaler()
X = scaler.fit_transform(X)
X_2d = scaler.fit_transform(X_2d)
C_range = np.logspace(-2, 10, 13)
gamma_range = np.logspace(-9, 3, 13)
param_grid = dict(gamma=gamma_range, C=C_range)
cv = StratifiedShuffleSplit(n_splits=5, test_size=0.2, random_state=42)
grid = GridSearchCV(SVC(), param_grid=param_grid, cv=cv)
grid.fit(X, y)
print("The best parameters are %s with a score of %0.2f    % (grid.best_params_, grid.best_score_))
C_2d_range = [1e-2, 1, 1e2]
gamma_2d_range = [1e-1, 1, 1e1]
classifiers = []
 for C in C_2d_range:
  for gamma in gamma_2d_range:
   clf = SVC(C=C, gamma=gamma)
clf.fit(X_2d, y_2d)
classifiers.append((C, gamma, clf))
plt.figure(figsize=(8, 6))

import  numpy as np
import  matplotlib.pyplot as plt
from  sklearn import svm, datasets
from  sklearn import model_selection
from  sklearn.metrics import classification_report
from  sklearn.metrics import confusion_matrix
iris = datasets.load_iris()
X = iris.data[:, :2]
y = iris.target
C = 1.0
svc = svm.SVC(kernel ='poly', C = 1).fit(X, y)
x_min, x_max = X[:, 0].min() -1, X[:, 0].max() + 1
y_min, y_max = X[:, 1].min() -1, X[:, 1].max() + 1
h = (x_max - x_min)/100
xx, yy = np.meshgrid(np.arange(x_min, x_max, h), np.arange(y_min, y_max, h))
plt.subplot(1, 1, 1)
Z = svc.predict(np.c_[xx.ravel(), yy.ravel()])
Z = Z.reshape(xx.shape)
plt.contourf(xx, yy, Z, cmap = plt.cm.Paired, alpha = 0.8)
plt.scatter(X[:, 0], X[:, 1], c = y, cmap = plt.cm.Paired)
plt.xlabel('Sepal length')
plt.ylabel('Sepal width')
plt.xlim(xx.min(), xx.max())
plt.title('SVC with poly kernel')
predicted = svc.predict(X)
matrix = confusion_matrix(y,predicted)
print(matrix)
report = classification_report(y,predicted)
print(report)
result=svc.score(X,y)
print("accuracy:%.3f%%"%(result*100.0))
plt.show()
```

Fig. 3 RBF kernel implementation using Python

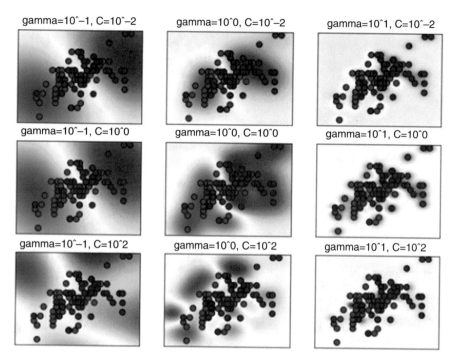

Fig. 4 RBF kernel output distribution

Linear kernel is indeed very well suited for text categorization. However [6], that is not the only solution and, in some cases, using another kernel might be better. Experimental results on iris dataset show that the approximate RBF-kernel SVM achieved classification performance and cross validation.

2   Kernel Ridge Regression

The duality relationship between ridge and [7] counter-ridge leads the way to kernels. If the data is relatively linear, linear regression/least squares are used to model the relationship between weight and size. So, fitting a line using least squares will minimize the sum of square residuals that ultimately gives size = size-intercept + (slope x weight), but this only works when there are a lot of measurements. If there are only a few measurements, for example, two measurements for training the model, then this leads to "over fitting" for the training data since the minimum sum of squared residuals is equal to 0 and variance is high to the testing data. This can be overcome using "ridge regression." Ridge regression fits a new line by introducing a small bias, and thus a considerable drop in variance can be seen in Fig. 8a and 8b.

In order to get better long-term predictions, this type [8] of regression can play a key role. In the equation size = y-intercept + slope x weight, the least squares are

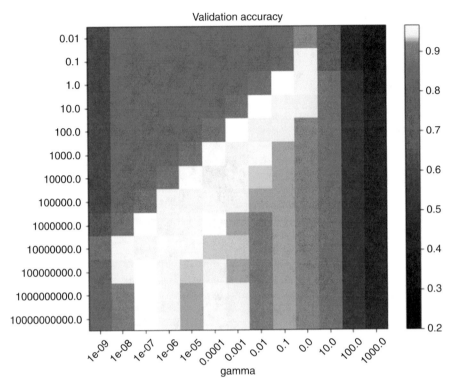

Fig. 5 RBF kernel accuracy

used to determine the values for the parameters. The sum of the squared residuals are also minimized + lambda x (slope)^2, the slope adds a penalty to the traditional least square method, and lambda determines how severe the penalty is.

In some cases, it is computationally overhead to calculate the lambda [9] and slope especially if the dimensions are regression-less and non-competent. In such cases, kernels are used to do the heavy work.

Advantages

1. It might be computationally efficient in some cases when solving [10] the system of equations.
2. By defining $K = XX^T$, we can work directly with K and never have to worry about X. This is the kernel trick.
3. Working with α is sometimes advantageous (e.g., in SVMs many entries of α will be zero).

Steps Involved

1. Initialize population and alpha
2. Identify A, b, and X such that $Ax = b$, where A is a feature set and b is the target at tribute and x is the relationship between A and b

```python
import  numpy as np
import  matplotlib.pyplot as plt
from  sklearn import svm, datasets
from  sklearn import model_selection
from  sklearn.metrics import classification_report
from  sklearn.metrics import confusion_matrix
iris = datasets.load_iris()
X = iris.data[:, :2]
y = iris.target
C = 1.0
svc = svm.SVC(kernel ='poly', C = 1).fit(X, y)
x_min, x_max = X[:, 0].min() -1, X[:, 0].max() + 1
y_min, y_max = X[:, 1].min() -1, X[:, 1].max() + 1
h = (x_max - x_min)/100
xx, yy = np.meshgrid(np.arange(x_min, x_max, h), np.arange(y_min, y_max, h))
plt.subplot(1, 1, 1)
Z = svc.predict(np.c_[xx.ravel(), yy.ravel()])
Z = Z.reshape(xx.shape)
plt.contourf(xx, yy, Z, cmap = plt.cm.Paired, alpha = 0.8)
plt.scatter(X[:, 0], X[:, 1], c = y, cmap = plt.cm.Paired)
plt.xlabel('Sepal length')
plt.ylabel('Sepal width')
plt.xlim(xx.min(), xx.max())
plt.title('SVC with poly kernel')
predicted = svc.predict(X)
matrix = confusion_matrix(y,predicted)
print(matrix)
report = classification_report(y,predicted)
print(report)
result=svc.score(X,y)
print("accuracy:%.3f%%"%(result*100.0))
plt.show()
```

Fig. 6 Code snippet of polynomial kernel using Python

Fig. 7 SVC with ploy kernel

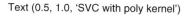
Text (0.5, 1.0, 'SVC with poly kernel')

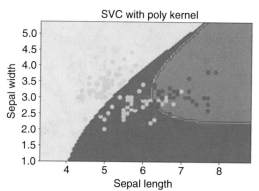

3. while (X==uniform or number of iterations)
4.   Calculate a regularization factor from the Tikhonov matrix (T = alpha x I)
5.   if T = 0 then calculate alpha from $A^T A$, alpha = $\|A^T A\|$
6.   Calculate x, x = $(A^T A + T^T T)^{-1} A^T b$
7.   update alpha, alpha = $\|Tx\|$
8. end while

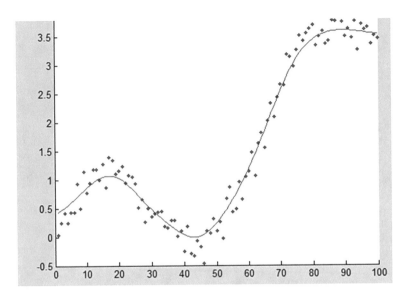

Fig. 8a Kernel ridge regression

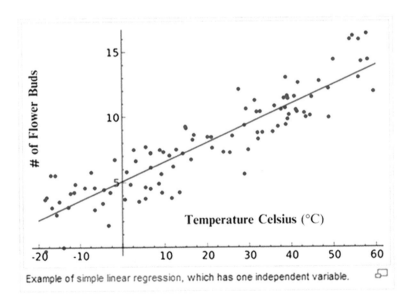

Fig. 8b Kernel ridge simple linear regression

Figure 9 provides the code implementation of the above-mentioned algorithm.

Figure 10 shows the output plot of the Python code implementation.

Figure 11 elucidates the Python code implementation [11] of Kernel Ridge Regression (KRR) prediction.

```
import numpy as np

rng = np.random.RandomState(0)

# Generate sample data
X = 5 * rng.rand(10000, 1)
y = np.sin(X).ravel()

# Add noise to targets
y[::5] += 3 * (0.5 - rng.rand(X.shape[0] // 5))

from sklearn.model_selection import GridSearchCV
from sklearn.kernel_ridge import KernelRidge
train_size = 100
kr = GridSearchCV(KernelRidge(kernel='rbf', gamma=0.1),param_grid={"alpha": [1e0, 0.1, 1e-2, 1e-3],
                              "gamma": np.logspace(-2, 2, 5)})
kr.fit(X[:train_size], y[:train_size])
X_plot = np.linspace(0, 5, 100000)[:, None]
y_kr = kr.predict(X_plot)

from sklearn.model_selection import learning_curve
import matplotlib.pyplot as plt

colors = np.random.random((3, 4))
plt.scatter(X[:100], y[:100], c='k', edgecolors=colors)
plt.plot(X_plot,y_kr, c='r',label="KRR fit")

plt.xlabel('data')
plt.ylabel('target')
plt.title('Kernel Ridge')
plt.legend()
plt.figure(figsize=(8, 6))
plt.show()
```

Fig. 9 Code implementation of aforementioned algorithm

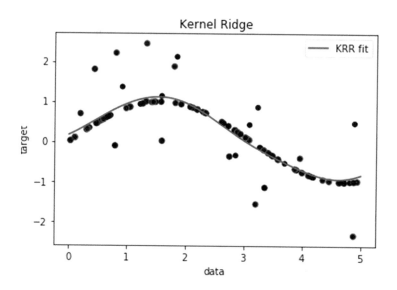

```
Figure size 576x432 with 0 Axes>
```

Fig. 10 Kernel ridge

```
import time
t0 = time.time()
kr.fit(X[:train_size], y[:train_size])
kr_fit = time.time() - t0
print("KRR complexity and bandwidth selected and model fitted in %.3f s"
      % kr_fit)

t0 = time.time()
y_kr = kr.predict(X_plot)
kr_predict = time.time() - t0
print("KRR prediction for %d inputs in %.3f s"
      % (X_plot.shape[0], kr_predict))
```

Fig. 11 Code snippet of KRR prediction

Output

KRR complexity and bandwidth selected and model fitted in 0.130 s.
KRR prediction for 100,000 inputs in 0.166 s.

We also need to formally optimize over λ. Specific λ choices, however, equate with specific B choices [12]. Using cross-validation or some other test, either λ or B should be chosen, so we can likewise fluctuate λ right now. In the ridge regression, there is no meaning for the vector support, which is one of a significant disadvantages in it. This is helpful in light of the fact that we possibly should summarize over the help vectors when [13] we test another model, which is a lot simpler than summarizing over the entire training set. In the SVM, the meager condition was resulting from the requirements of disparity in light of the fact that the integral states of slackness revealed to us that regardless of whether the limitation was idle, at that point the multiplier αi was zero. There is no impact of that type here.

3 Kernel Density Estimator

The Kernel is a non-negative, real-valued probability distribution [14] function with an even definite integral, which sets equal to a value of 1. For a continuous random variable, estimating the probability density function is done by the kernel density estimation.

Characteristics

1. Nonparametric technique
2. Effective multimodal data representation
3. Consideration of noise for observed data
4. Representation of model/state

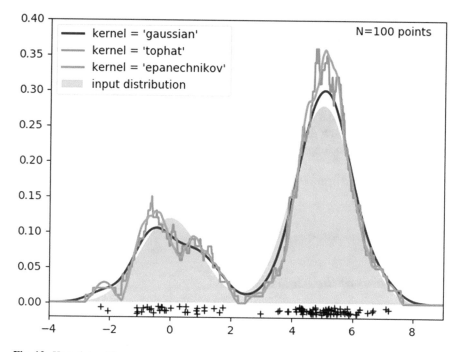

Fig. 12 Kernel densities

3.1 Density Estimation

The three different techniques, unsupervised learning, data modeling, and feature engineering, are the ones required for the density estimation [15] and they have to walk through the line between them to get the result. In literature, many density estimation models are available, but Gaussian mixtures and neighbor-based models like kernel density estimation are more popular than the others.

In density estimation, visualization of information can be achieved using the histogram in which bins can be defined, but the problem is that choosing the [16] bins has a disproportionate effect on results in visualization, so kernel density estimator came to exist, which can be presented in several [17] numbers of dimensions and it uses ball tree or Kernel Density (KD) tree for efficient queries. Here kernel may be a Gaussian, Tophat, or Epanechnikov (Fig. 12).

3.2 Constructing a Kernel Density Estimate

1. Select kernel that performs efficiently for the given dataset.

2. On every datum (p_i), construct a scaled kernel function:

$$h^{-1}K\left[(p-p_i)\right]/h.$$

where k = chosen kernel function

h is a bandwidth, known as smoothing parameter also called as window width

3. Sum up the individual scaled kernel functions and divide by n, this places a probability of 1/n to each datum x_i. it also ensures the kernel density estimate integrates to 1 over its support set:

$$\bar{f}(x) = n^{-1}h^{-1}\sum_{i=1}^{n}K\left[\frac{p-p_i}{h}\right]$$

3.3 Features of the Algorithm

3.3.1 Bandwidth Selection

Bandwidth selection gives the optimal bandwidth for the solution and [18] it can estimates from reference rules (Silverman's rule) which less impact and another empirical approach is Cross validation is often used.

- Automatic data-dependent bandwidth selection (MISE – mean integrated squared error) – error between estimated and true error
- Variable bandwidth selection – mean shift which is not effective

For choosing bandwidth, small h results small standard deviation best suits when sample size is large and data are tightly packed. A large h results large standard deviation best suits when sample size is small, and data are sparse.

3.3.2 Kernels

KDE can be implemented using different kernels that lead to different characteristics of density estimates. As Scipy contains Gaussian, stats models have 7 kernels, and Scikit-learn have 6 and each uses a dozen distance metrics for different kernel shapes (Fig. 13).

Plot. Kernels ()

3.3.3 Heterogeneous Data

States models can be a heterogeneous data. In general, this [19] data is nothing but a combination of continuous, ordered, and unordered discrete variables.

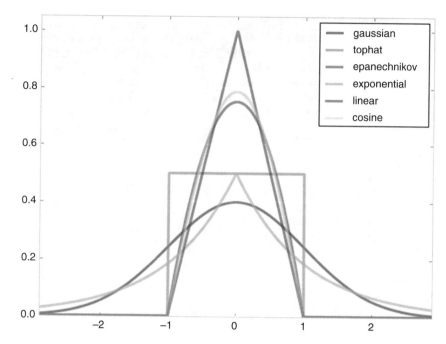

Fig. 13 Constructing different kernel level densities

3.3.4 Fast Fourier Transform–Based Computations

For large datasets, KDE is computed efficiently using fast Fourier transform (FFT) but requires binning and becomes inefficient in higher dimensions.

3.3.5 Tree-Based Computations

Using the KD tree that is a specialized data structure, c is used to compute the KDE in which it is required to compute M evaluations of N points.

3.3.6 Computational Efficiency

By comparing the computational efficiency of different algorithms (Table 1), it depends on the number of scaling points. It is better in one-dimensional data than in multi-dimensional data.

Scikit-learn computers faster than other implementation. But when data is heterogeneous, stat model is better and Scipy's Gaussian KDE is used to obtain the results (Figs. 14 and 15).

Table 1 Computational efficiency matrix

	Bandwidth selection	Available Kernels	Multi-dimension	Heterogeneous data	FFT-based computation	Tree-based computation
Scipy	Scott & Silverman	One (Gauss)	Yes	No	No	No
Stats models KDE Univariate	Scott & Silverman	Seven	1D only	No	Yes	No
Stats models KDE Multivariate	Normal reference cross-validation	Seven	Yes	Yes	No	No
Scikit-learn	None built-in; cross val. available	6 kernels x 12 metrics	Yes	No	No	Ball tree or KD tree

```
from scipy import stats
from scipy.stats.kde import gaussian_kde
from scipy.stats import norm
import numpy as np
import seaborn as sns
import matplotlib.pyplot as plt
from numpy import linspace,hstack
# creating data with two peaks
sampD1 = norm.rvs(loc=-1.0,scale=1,size=300)
sampD2 = norm.rvs(loc=2.0,scale=0.5,size=300)
rs = hstack([sampD1,sampD2])
# for the Guassian Kernel
kde = gaussian_kde(rs)
kde_low_bw = gaussian_kde(rs, bw_method=0.10)
x = np.linspace(rs.min()-0.01,rs.max()+0.01,200)
fig, axes = plt.subplots(1, 3, figsize=(15, 3))
axes[0].hist(rs,alpha=0.5,bins=25)
axes[1].plot(x,kde(x),label='KDE')
axes[1].plot(x,kde_low_bw(x),label='KDE(low bw)')
axes[1].legend()
sns.distplot(rs, fit=stats.norm, bins=25, ax=axes[2])    #to create different distributions
plt.show()[4]
#score_samples returns the log of the probability density
logprob = kde.score_samples (x[:,None])
Plt.fill_between (x,np.exp(logprob),alpha =0.5)
Plt.plot(x,np.full_like(x,-0.01,'|k\,markeredgewidth =1)
Print ("Log probability density value is :")
```

Fig. 14 Code implementation of KDE

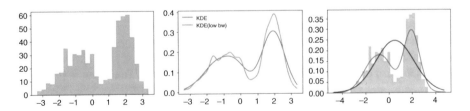

Fig. 15 Log of the probability density value is (−0.02,0.22)

Code Implementation

Advantages

When compared with the commonly used histogram, the kernel density [20] estimator shows several advantages.

1. It is a smooth curve and thus it better exhibits the details of the PDF suggesting in some cases non-unimodality.
2. It uses all sample points' locations, therefore, it better reveals the information contained in the sample.
3. It convincingly suggests multimodality.
4. The bias of the kernel estimator is of one order better than that of a histogram estimator.
5. Compared with 1D application, 2D kernel applications are even better as the 2D histogram.

Disadvantages

1. Annoying artifacts, such as all-positive quantities whose kernel density estimates go into the negative zone. This can be fixed, but (a) it typically is not, and (b) when there is no an obvious bound, you still have the issue of the kernel density including places.
2. PDF per pixel by KDE, classification by global threshold.
3. Computational cost is high.
4. Memory consumption.
5. Bandwidth selection issue.

Applications

1. Density level estimation.
2. Clustering or unsupervised learning.
3. Description of main content of data.

A range of kernel functions are commonly used: uniform, triangular, biweight, triweight, Epanechnikov, normal, and others. Based on the requirement of the result and the available dataset, the kernel estimator is used. For accurate results, Gaussian kernel is implemented.

4 Dimensionality Reduction with Kernel Principal Component Analysis

Principal component analysis (PCA) is a tool that is used to reduce the dimension of the [21] data. It allows reducing the dimension of the data without much loss of information. PCA reduces the dimension by finding a few orthogonal linear combinations (principal components) of the original variables with the largest variance. The first principal component captures most of the variance in the data. The [22] second principal component is orthogonal to the first principal component and captures the remaining variance. PCA is a linear method. That is, it can only be applied to datasets that are linearly separable. It does an excellent job for datasets, which are linearly separable. However, if we use it to non-linear datasets, we might get a result that may not be the optimal dimensionality reduction. Kernel PCA uses a kernel function to project dataset into a higher dimensional feature space, where it is linearly separable. Figures 16, 17, 18, 19, 20 and 21 illustrate the various PCA Python code implementations and their corresponding plots.

```python
import matplotlib.pyplot as plt
from sklearn.datasets import make_moons

X, y = make_moons(n_samples = 500, noise = 0.02, random_state = 417)

plt.scatter(X[:, 0], X[:, 1], c = y)
plt.show()
```

Fig. 16 Python code snippet for make_moons

Fig. 17 Make-moons plot

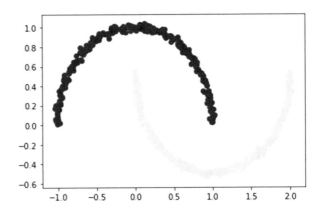

```python
from sklearn.decomposition import PCA
pca = PCA(n_components = 2)
X_pca = pca.fit_transform(X)

plt.title("PCA")
plt.scatter(X_pca[:, 0], X_pca[:, 1], c = y)
plt.xlabel("Component 1")
plt.ylabel("Component 2")
plt.show()
```

Fig. 18 PCA fit Python code snippet

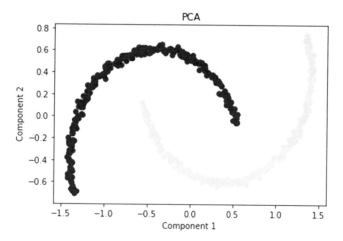

Fig. 19 PCA plot

```
from sklearn.decomposition import KernelPCA
kpca = KernelPCA(kernel ='rbf', gamma = 15)
X_kpca = kpca.fit_transform(X)

plt.title("Kernel PCA")
plt.scatter(X_kpca[:, 0], X_kpca[:, 1], c = y)
plt.show()
```

Fig. 20 Kernel PCA implementation in Python

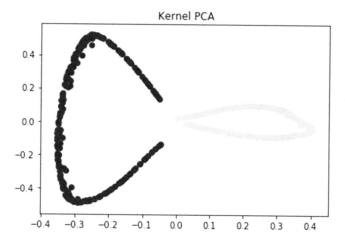

Fig. 21 Kernel PCA slot

Algorithm

1. Construct the covariance matrix of the data.
2. Compute the eigenvectors of this matrix.
3. Eigenvectors corresponding to the largest eigenvalues are used to reconstruct a large fraction of variance of the original data.
4. Hence, we are left with a lesser number of eigenvectors, and there might have been some data loss in the process.

Code Implementations

Advantages

1. It helps in data compression, and hence reduced storage space.
2. It reduces computation time.
3. It also helps remove redundant features, if any.

Disadvantages

1. It may lead to some amount of data loss.
2. PCA tends to find linear correlations between variables, which is sometimes undesirable.
3. PCA fails in cases where mean and covariance are not enough to define datasets.
4. We may not know how many principal components to keep in practice, and some thumb rules are applied.

5   Hidden Markov Model to Estimate the Behavior of a Person or Animal

A Markov chain (model) describes a stochastic process where [23] the assumed probability of future state(s) depends only on the current process state and not on any of the states that preceded it. The hidden Markov model (HMM) is a machine learning algorithm that is part of the graphic models. Nevertheless, HMM is often trained using a supervised method of learning in the [24] case of data being available for training. Only a little bit of probability knowledge will suffice to understand the concept to anyone.

It is important to understand where the HMM algorithm is used. In short, HMM is a graphical model, which is generally used in predicting states (hidden) using sequential data like weather, text, speech, etc. Figures 22, 23, 24, 25, 26 and 27 elucidate HMM that estimates the behavior of a person or animal.

Algorithm

1. Import library packages NumPy, pandas.
2. Import HMM library package.

```python
import numpy as np
import pandas as pd

obs_map = {'eating':0, 'sleeping':1}
obs = np.array([1,1,0,1,0,0,1,0,1,1,0,0,0,1])

inv_obs_map = dict((v,k) for k, v in obs_map.items())
obs_seq = [inv_obs_map[v] for v in list(obs)]

print("Simulated Observations:\n",pd.DataFrame(np.column_stack([obs, obs_seq]),
        columns=['Obs_code', 'Obs_seq']) )

pi = [0.6,0.4]
states = ['eating', 'sleeping']
hidden_states = ['healthy', 'sick', 'normal']
pi = [0, 0.2, 0.8]
state_space = pd.Series(pi, index=hidden_states, name='states')
a_df = pd.DataFrame(columns=hidden_states, index=hidden_states)
a_df.loc[hidden_states[0]] = [0.3, 0.3, 0.4]
a_df.loc[hidden_states[1]] = [0.1, 0.45, 0.45]
a_df.loc[hidden_states[2]] = [0.2, 0.3, 0.5]
print("\n HMM matrix:\n", a_df)
a = a_df.values

observable_states = states
b_df = pd.DataFrame(columns=observable_states, index=hidden_states)
b_df.loc[hidden_states[0]] = [1,0]
b_df.loc[hidden_states[1]] = [0.8,0.2]
b_df.loc[hidden_states[2]] = [0.3,0.7]
print("\n Observable layer  matrix:\n",b_df)
b = b_df.values
```

Fig. 22 HMM Python code snippet

3. Init blank path.
4. The forward algorithm extension.
5. Find optimal path.

Code Implementation

Advantages

1. Statistical base of HMM is strong.
2. Efficient learning algorithm can take place directly [25] from raw sequence of data.
3. It has a wide variety of applications like data mining classifications, structural analysis, and pattern discovery.

Disadvantages

1. HMM often has a large number of unstructured parameters.

```
Simulated Observations:
    Obs_code   Obs_seq
0          1   sleeping
1          1   sleeping
2          0     eating
3          1   sleeping
4          0     eating
5          0     eating
6          1   sleeping
7          0     eating
8          1   sleeping
9          1   sleeping
10         0     eating
11         0     eating
12         0     eating
13         1   sleeping

HMM matrix:
          healthy   sick  normal
healthy      0.3    0.3     0.4
sick         0.1    0.45    0.45
normal       0.2    0.3     0.5

Observable layer  matrix:
          eating sleeping
healthy      1        0
sick         0.8      0.2
normal       0.3      0.7
```

Fig. 23 Output of HMM

```
def viterbi(pi, a, b, obs):

    nStates = np.shape(b)[0]
    T = np.shape(obs)[0]

    # init blank path
    path = path = np.zeros(T,dtype=int)
    # delta --> highest probability of any path that reaches state i
    delta = np.zeros((nStates, T))
    # phi --> argmax by time step for each state
    phi = np.zeros((nStates, T))

    # init delta and phi
    delta[:, 0] = pi * b[:, obs[0]]
    phi[:, 0] = 0

    print('\nStart Walk Forward\n')
    # the forward algorithm extension
    for t in range(1, T):
        for s in range(nStates):
            delta[s, t] = np.max(delta[:, t-1] * a[:, s]) * b[s, obs[t]]
            phi[s, t] = np.argmax(delta[:, t-1] * a[:, s])
            print('s={s} and t={t}: phi[{s}, {t}] = {phi}'.format(s=s, t=t, phi=phi[s, t]))
```

Fig. 24 HMM code implementation part 1

```
# find optimal path
print('-'*50)
print('Start Backtrace\n')
path[T-1] = np.argmax(delta[:, T-1])
for t in range(T-2, -1, -1):
    path[t] = phi[path[t+1], [t+1]]
    print('path[{}] = {}'.format(t, path[t]))

return path, delta, phi
```

```
path, delta, phi = viterbi(pi, a, b, obs)
state_map = {0:'healthy', 1:'sick', 2:'normal'}
state_path = [state_map[v] for v in path]
pd.DataFrame().assign(Observation=obs_seq).assign(Best_Path=state_path)
```

Fig. 25 HMM code implementation part 2

2. First-order HMM is limited by its first order of Markov property.
3. They cannot express dependencies between hidden states.
4. Modeling protein folds into a complex 3D shape determining its function.

Applications

1. Gene prediction.
2. Modeling protein domains.
3. Clustering of paths for a subgroup.

HMMs are used in a variety of scenarios including manipulation of the natural language, robotics, and biogenetics. We have seen some of the basics of HMMs in this section, particularly in the context estimating the behavior of a person or an animal.

6 Factor Analysis

Factor analysis is one of the statistical methods that measures of how much one observed and correlated variable [26] vary with another by a set of less or unobserved variables. In other words, it simply is the method that defines the covariance relationship between the set of observed variables.

For instance, some data on particular set of people of observed characteristics of people. For example, do people having *insomnia* have *suicidal thoughts* or feel *nauseous* most of the time and (say) have *covariance (insomnia, suicidal thoughts) = 0.2*. Factor analysis works [27] by supposing that variance and covariance structure in the observed characteristics is due to unobserved factors (say) such as *depression*

```
Start Walk Forward

s=0 and t=1: phi[0, 1] = 2.0
s=1 and t=1: phi[1, 1] = 2.0
s=2 and t=1: phi[2, 1] = 2.0
s=0 and t=2: phi[0, 2] = 2.0
s=1 and t=2: phi[1, 2] = 2.0
s=2 and t=2: phi[2, 2] = 2.0
s=0 and t=3: phi[0, 3] = 0.0
s=1 and t=3: phi[1, 3] = 1.0
s=2 and t=3: phi[2, 3] = 1.0
s=0 and t=4: phi[0, 4] = 2.0
s=1 and t=4: phi[1, 4] = 2.0
s=2 and t=4: phi[2, 4] = 2.0
s=0 and t=5: phi[0, 5] = 0.0
s=1 and t=5: phi[1, 5] = 1.0
s=2 and t=5: phi[2, 5] = 1.0
s=0 and t=6: phi[0, 6] = 0.0
s=1 and t=6: phi[1, 6] = 1.0
s=2 and t=6: phi[2, 6] = 1.0
s=0 and t=7: phi[0, 7] = 2.0
s=1 and t=7: phi[1, 7] = 2.0
s=2 and t=7: phi[2, 7] = 2.0
s=0 and t=8: phi[0, 8] = 0.0
s=1 and t=8: phi[1, 8] = 1.0
s=2 and t=8: phi[2, 8] = 1.0
s=0 and t=9: phi[0, 9] = 2.0
s=1 and t=9: phi[1, 9] = 2.0
s=2 and t=9: phi[2, 9] = 2.0
s=0 and t=10: phi[0, 10] = 2.0
s=1 and t=10: phi[1, 10] = 2.0
s=2 and t=10: phi[2, 10] = 2.0
s=0 and t=11: phi[0, 11] = 0.0
s=1 and t=11: phi[1, 11] = 1.0
s=2 and t=11: phi[2, 11] = 1.0
s=0 and t=12: phi[0, 12] = 0.0
s=1 and t=12: phi[1, 12] = 1.0
s=2 and t=12: phi[2, 12] = 1.0
s=0 and t=13: phi[0, 13] = 0.0
s=1 and t=13: phi[1, 13] = 1.0
s=2 and t=13: phi[2, 13] = 1.0
```

Fig. 26 HMM code results part 1

responsible for the variance between all of the other observed variables. It simply describes the variance and covariance by supposing a casual effect of the unobserved underlying factors on the observed characteristics.

Factor analysis is an extension of PCA. Both models try to approximate the covariance matrix Σ, but factor analysis questions whether the data are consistent with some prescribed structure. Figures 28, 29, 30, 31, 32, 33, 34, 35, 36 and 37 describe the step-wise implementation in Python with various algorithms.

Code Implementation

```
--------------------------------
Start Backtrace

path[12] = 1
path[11] = 1
path[10] = 1
path[9] = 2
path[8] = 2
path[7] = 1
path[6] = 2
path[5] = 1
path[4] = 1
path[3] = 2
path[2] = 1
path[1] = 2
path[0] = 2
```

Fig. 26 (continued)

Fig. 27 HMM code results part 2

	Observation	Best_Path
0	sleeping	normal
1	sleeping	normal
2	eating	sick
3	sleeping	normal
4	eating	sick
5	eating	sick
6	sleeping	normal
7	eating	sick
8	sleeping	normal
9	sleeping	normal
10	eating	sick
11	eating	sick
12	eating	sick
13	sleeping	normal

Factor 1 has high factor loadings for E1, E2, E3, E4, and E5 (extraversion).
Factor 2 has high factor loadings for N1, N2, N3, N4, and N5 (neuroticism).
Factor 3 has high factor loadings for C1, C2, C3, C4, and C5 (conscientiousness).
Factor 4 has high factor loadings for O1, O2, O3, O4, and O5 (openness).
Factor 5 has high factor loadings for A1, A2, A3, A4, and A5 (agreeableness).
Factor 6 has none of the high loadings for any variable and is not easily interpretable. It is good if we take only five factors (Fig. 38). Figures 39 and 40 show the Python code implementation and variance results.

```
!pip install factor_analyzer==0.2.3
# Import required Libraries
import pandas as pd
from sklearn.datasets import load_iris
from factor_analyzer import FactorAnalyzer
import matplotlib.pyplot as plt
```

```
df= pd.read_csv("bfi.csv")
```

Preprocessing

```
df.columns
```

```
Index(['Unnamed: 0', 'A1', 'A2', 'A3', 'A4', 'A5', 'C1', 'C2', 'C3', 'C4',
       'C5', 'E1', 'E2', 'E3', 'E4', 'E5', 'N1', 'N2', 'N3', 'N4', 'N5', 'O1',
       'O2', 'O3', 'O4', 'O5', 'gender', 'education', 'age'],
      dtype='object')
```

```
# Dropping unnecessary columns
df.drop(['gender', 'education', 'age','Unnamed: 0'],axis=1,inplace=True)
```

```
# Dropping missing values rows
df.dropna(inplace=True)
```

Fig. 28 Factor analysis Python code implementation

7   Twitter Sentiment Analysis

Required Packages

- Tweepy (!pip install tweepy)
- NumPy (!pip install numpy)
- Pandas (!pip install pandas)
- Matplotlib.pyplot (!pip install matplotlib.pyplot)
- Json (!pip install json)
- Textblob (!pip install textblob)
- Re (!pip install re)

To do Twitter analysis, we need ACCESS_TOKEN, ACCESS_TOKEN_SECRET, CONSUMER_KEY, and CONSUMER_SECRET, which can be generated from the Twitter developer dashboard. If you are note a Twitter [28] developer, visit https://developer.twitter.com/ and register with a Twitter account. In order to become a Twitter developer, you need to mention why you would like to become a Twitter

```
df.info()
```

```
<class 'pandas.core.frame.DataFrame'>
Int64Index: 2436 entries, 0 to 2799
Data columns (total 25 columns):
A1    2436 non-null float64
A2    2436 non-null float64
A3    2436 non-null float64
A4    2436 non-null float64
A5    2436 non-null float64
C1    2436 non-null float64
C2    2436 non-null float64
C3    2436 non-null float64
C4    2436 non-null float64
C5    2436 non-null float64
E1    2436 non-null float64
E2    2436 non-null float64
E3    2436 non-null float64
E4    2436 non-null float64
E5    2436 non-null float64
N1    2436 non-null float64
N2    2436 non-null float64
N3    2436 non-null float64
N4    2436 non-null float64
N5    2436 non-null float64
O1    2436 non-null float64
O2    2436 non-null int64
O3    2436 non-null float64
O4    2436 non-null float64
O5    2436 non-null float64
dtypes: float64(24), int64(1)
memory usage: 494.8 KB
```

Fig. 29 Factor analysis Python implementation part 1

```
#checking factorability by bartlett's test method
from factor_analyzer.factor_analyzer import calculate_bartlett_sphericity
chi_square_value,p_value=calculate_bartlett_sphericity(df)
chi_square_value, p_value
```

Fig. 30 Factor analysis Python implementation Bartlett's test method part 2

```
#using kaiser-meyer-olkin test
from factor_analyzer.factor_analyzer import calculate_kmo
kmo_all,kmo_model=calculate_kmo(df)
```

```
kmo_model
```

```
0.848539722194922
```

Fig. 31 Factor analysis part 3

```
""" choosing number of factors """
# Create factor analysis object and perform factor analysis
fa = FactorAnalyzer(rotation=None)
fa.fit(df, 25)
# Check Eigenvalues
ev, v = fa.get_eigenvalues()
pd.DataFrame(ev)
```

Fig. 32 Factor analysis part 4

Fig. 33 Factor analysis part 5

	0		
0	5.134311		
1	2.751887		
2	2.142702		
3	1.852328		
4	1.548163		
5	1.073582	15	0.543305
6	0.839539	16	0.514518
7	0.799206	17	0.494503
8	0.718989	18	0.482640
9	0.688089	19	0.448921
10	0.676373	20	0.423366
11	0.651800	21	0.400671
12	0.623253	22	0.387804
13	0.596563	23	0.381857
14	0.563091	24	0.262539

```
# Create scree plot using matplotlib
plt.scatter(range(1,df.shape[1]+1),ev)
plt.plot(range(1,df.shape[1]+1),ev)
plt.title('Scree Plot')
plt.xlabel('Factors')
plt.ylabel('Eigenvalue')
plt.grid()
plt.show()
```

Fig. 34 Factor analysis part 6

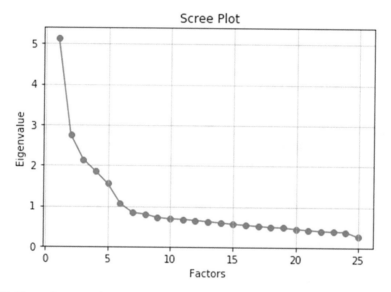

Fig. 35 Eigenvalue scree plot

developer and what kind of data you are going to access and where are you going to implement the data. Once you are a developer, create an app by mentioning the app name, application description, website URL (Github repository link is acceptable), callback URL (localhost / http:127.0.0.1:8080), and your usage. Once you finish creating an app go to https://developer.twitter.com/en/apps and navigate to the details of your app to access keys and tokens where you will find consumer keys and access keys.

Now it is time to install packages. After installing all the packages create a Python file.

In that file,

1. Import packages.
2. Declare ACCESS_TOKEN, ACCESS_TOKEN_SECRET, CONSUMER_KEY, and CONSUMER_SECRET variables and assign your respective keys.

```
# Create scree plot using matplotlib
plt.scatter(range(1,df.shape[1]+1),ev)
plt.plot(range(1,df.shape[1]+1),ev)
plt.title('Scree Plot')
plt.xlabel('Factors')
plt.ylabel('Eigenvalue')
plt.grid()
plt.show()
```

```
# Create factor analysis object and perform factor analysis
fa = FactorAnalyzer(rotation="varimax",n_factors=6)
fa.fit(df, 6)
```

```
FactorAnalyzer(bounds=(0.005, 1), impute='median', is_corr_matrix=False,
        method='minres', n_factors=6, rotation='varimax',
        rotation_kwargs={}, use_smc=True)
```

Fig. 36 Factor analyzer scree plot

3. Authenticate your consumer keys with twitter by passing consumer keys to the tweepy.OAuthHandler method from the tweepy package that we have imported earlier.
4. Pass access tokens to the set_access_token() method of the authenticated Twitter object.
5. Get your API by passing the authenticated object to API() from the tweepy package.

Code

```
ACCESS_TOKEN = "your access token"
ACCESS_TOKEN_SECRET = "your access token secret"
CONSUMER_KEY = "your consumer key"
CONSUMER_SECRET = "your consumer secret"
auth = tweepy.OAuthHandler(CONSUMER_KEY,CONSUMER_SECRET)
auth.set_access_token(ACCESS_TOKEN,ACCESS_TOKEN_SECRET)
api = tweepy.API(auth)
```

Now that we have access to twitter API, we can use that API to stream data. Streaming data from twitter can be possible in 2 ways:

1. Using the tweepy API calling method
2. Using the tweepy cursor, streamListener, and a little OOPS implementation

Using the tweepy API Calling Method
tweepy.api class provides various wrappers for API provided by Twitter. Such wrappers include many Timeline Methods:

```
pd.DataFrame(fa.loadings_)
```

	0	1	2	3	4	5
0	0.095220	0.040783	0.048734	-0.530987	-0.113057	0.161216
1	0.033131	0.235538	0.133714	0.661141	0.063734	-0.006244
2	-0.009621	0.343008	0.121353	0.605933	0.033990	0.160106
3	-0.081518	0.219717	0.235140	0.404594	-0.125338	0.086356
4	-0.149616	0.414458	0.106382	0.469698	0.030977	0.236519
5	-0.004358	0.077248	0.554582	0.007511	0.190124	0.095035
6	0.068330	0.038370	0.674545	0.057055	0.087593	0.152775
7	-0.039994	0.031867	0.551164	0.101282	-0.011338	0.008996
8	0.216283	-0.066241	-0.638475	-0.102617	-0.143846	0.318359
9	0.284187	-0.180812	-0.544838	-0.059955	0.025837	0.132423
10	0.022280	-0.590451	0.053915	-0.130851	-0.071205	0.156583
11	0.233624	-0.684578	-0.088497	-0.116716	-0.045561	0.115065
12	-0.000895	0.556774	0.103390	0.179396	0.241180	0.267291
13	-0.136788	0.658395	0.113798	0.241143	-0.107808	0.158513
14	0.034490	0.507535	0.309813	0.078804	0.200821	0.008747
15	0.805806	0.068011	-0.051264	-0.174849	-0.074977	-0.096266
16	0.789832	0.022958	-0.037477	-0.141134	0.006726	-0.139823
17	0.725081	-0.065687	-0.059039	-0.019184	-0.010664	0.062495
18	0.578319	-0.345072	-0.162174	0.000403	0.062916	0.147551
19	0.523097	-0.161675	-0.025305	0.090125	-0.161892	0.120049
20	-0.020004	0.225339	0.133201	0.005178	0.479477	0.218690
21	0.156230	-0.001982	-0.086047	0.043989	-0.496640	0.134693
22	0.011851	0.325954	0.093880	0.076642	0.566128	0.210777
23	0.207281	-0.177746	-0.005671	0.133656	0.349227	0.178068
24	0.063234	-0.014221	-0.047059	-0.057561	-0.576743	0.135936

Fig. 37 Output-factor analysis part 1

```
# Create factor analysis object and perform factor analysis using 5 factors
fa = FactorAnalyzer(rotation="varimax",n_factors=5)
fa.fit(df, 5)
pd.DataFrame(fa.loadings_)
```

Fig. 38 Factor analysis fitting

	0	1	2	3	4
0	0.111126	0.040465	0.022798	-0.428166	-0.077931
1	0.029588	0.213716	0.139037	0.626946	0.062139
2	0.009357	0.317848	0.109331	0.650743	0.056196
3	-0.066476	0.204566	0.230584	0.435624	-0.112700
4	-0.122113	0.393034	0.087869	0.537087	0.066708
5	0.010416	0.070184	0.545824	0.038878	0.209584
6	0.089574	0.033270	0.648731	0.102782	0.115434
7	-0.030855	0.023907	0.557036	0.111578	-0.005183
8	0.240410	-0.064984	-0.633806	-0.037498	-0.107535
9	0.290318	-0.176395	-0.562467	-0.047525	0.036822
10	0.042819	-0.574835	0.033144	-0.104813	-0.058795
11	0.244743	-0.678731	-0.102483	-0.112517	-0.042010
12	0.024180	0.536816	0.083010	0.257906	0.280877
13	-0.115614	0.646833	0.102023	0.306101	-0.073422
14	0.036145	0.504069	0.312899	0.090354	0.213739
15	0.786807	0.078923	-0.045997	-0.216363	-0.084704
16	0.754109	0.027301	-0.030568	-0.193744	-0.010304
17	0.731721	-0.061430	-0.067084	-0.027712	-0.004217
18	0.590602	-0.345388	-0.178902	0.005886	0.075225
19	0.537858	-0.161291	-0.037309	0.100931	-0.149769
20	-0.002224	0.213005	0.115080	0.061550	0.504907

	0	1	2	3	4	
13	-0.136788	0.658395	0.113798	0.241143	-0.107808	0.158513
14	0.034490	0.507535	0.309813	0.078804	0.200821	0.008747
15	0.805806	0.068011	-0.051264	-0.174849	-0.074977	-0.096266
16	0.789832	0.022958	-0.037477	-0.141134	0.006726	-0.139823
17	0.725081	-0.065687	-0.059039	-0.019184	-0.010664	0.062495
18	0.578319	-0.345072	-0.162174	0.000403	0.062916	0.147551
19	0.523097	-0.161675	-0.025305	0.090125	-0.161892	0.120049
20	-0.020004	0.225339	0.133201	0.005178	0.479477	0.218690
21	0.156230	-0.001982	-0.086047	0.043989	-0.496640	0.134693
22	0.011851	0.325954	0.093880	0.076642	0.566128	0.210777
23	0.207281	-0.177746	-0.005671	0.133656	0.349227	0.178068
24	0.063234	-0.014221	-0.047059	-0.057561	-0.576743	0.135936

Fig. 39 Output-factor analysis part 2

```
# Get variance of each factors
fa.get_factor_variance()
```

```
(array([2.70963262, 2.47308983, 2.04110564, 1.844498  , 1.52215297]),
 array([0.1083853 , 0.09892359, 0.08164423, 0.07377992, 0.06088612]),
 array([0.1083853 , 0.2073089 , 0.28895312, 0.36273304, 0.42361916]))
```

```
print("Variance percentage:",fa.get_factor_variance()[-1][-1]*100)
```

```
Variance percentage: 42.36191619470739
```

Fig. 40 Variance percentage output

```
API.home_timeline([since_id][, max_id][, count][, page])
API.statuses_lookup(id[, include_entities][, trim_user][, map])
API.user_timeline([id/user_id/screen_name][, since_id][, max_id]
[, count][, page])
API.retweets_of_me([since_id][, max_id][, count][, page])
```

For many more, please refer to the following: http://docs.tweepy.org/en/v3.5.0/api.html

```
tweets = api.user_timeline(screen_name='cmanmohan', count=20)
```

Tweets variable contains a list of 20 recent tweets from the user "cmanmohan" as a dictionary. For sentimental analysis we are considering the tweet text from all those tweets, so we need to extract the data from the "text" directory from each dictionary consisting of ['__class__', '__delattr__', '__dict__', '__dir__', '__doc__', '__eq__', '__format__', '__ge__', '__getattribute__', '__getstate__', '__gt__', '__hash__', '__init__', '__init_subclass__', '__le__', '__lt__', '__module__', '__ne__', '__new__', '__reduce__', '__reduce_ex__', '__repr__', '__setattr__', '__sizeof__', '__str__', '__subclasshook__', '__weakref__', '_api', '_json', 'author', 'contributors', 'coordinates', 'created_at', 'destroy', 'entities', 'favorite', 'favorite_count', 'favorited', 'geo', 'id', 'id_str', 'in_reply_to_screen_name', 'in_reply_to_status_id', 'in_reply_to_status_id_str', 'in_reply_to_user_id', 'in_reply_to_user_id_str', 'is_quote_status', 'lang', 'parse', 'parse_list', 'place', 'possibly_sensitive', 'quoted_status', 'quoted_status_id', 'quoted_status_id_str', 'retweet', 'retweet_count', 'retweeted', 'retweets', 'source', 'source_url', 'text', 'truncated', 'user'] directories and include those into a pandas Data Frame for processing.

```
df = pd.DataFrame(data = [tweet.text for tweet in tweets],
columns=['Tweets'])
```

Additionally, we can include other directory data like retweet_count, favorite_count, and created_at for visualization purposes (Fig. 41):

	Tweets	retweets	likes	date
0	RT @DPP_PKB: Bersatu Menghadapi Endemik Virus ...	0	0	Wed Jan 29 09:47:57 +0000 2020
1	RT @Pierrefirdaus: 'Vaksin' yang paling kuat a...	0	0	Wed Jan 29 09:47:57 +0000 2020
2	RT @xicograziano: Cara, esse Zé de Abreu tem u...	0	0	Wed Jan 29 09:47:57 +0000 2020
3	RT @parcha_p: งานเข้าแล้วมั้ยคะ เมื่อไทยถูกหมอ...	0	0	Wed Jan 29 09:47:57 +0000 2020
4	Bueno queda cancelada mi experiencia este vera...	0	0	Wed Jan 29 09:47:57 +0000 2020

Fig. 41 Tweet retweets and likes tables from twitter

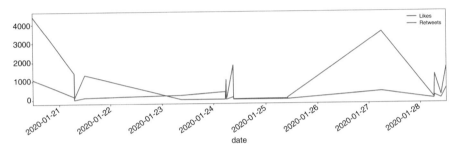

Fig. 42 Plot the likes and retweets of sentiment analysis

```
df['retweets']  =  np.array([tweet.retweet_count  for  tweet  in
tweets])
df['likes'] = np.array([tweet.favorite_count for tweet in tweets])
df['date'] = np.array([tweet.created_at for tweet in tweets])
df.head()
```

With the data we have in df Data Frame, we can visualize a plot as shown in Fig. 42 that illustrates likes and retweets against dates:

```
# time likes
time_retweets           =           pd.Series(data=df['likes'].values,
index=df['date'])
time_retweets.plot(figsize=(16,4),color='b',legend=True,label="Li
kes")
# timeretweets
time_retweets           =           pd.Series(data=df['retweets'].values,
index=df['date'])
time_retweets.plot(figsize=(16,4),color='r',legend=True,label="Ret
weets")
plt.show()
```

Coming to our main objective (sentiment analysis), we know that our data frame has Tweets, retweets, likes, and date as attributes and [29] that out target attribute is Tweets that contain hyperlinks, special characters, and other noisy elements, and hence we need to clean the strings before proceeding to analysis. There are several ways of cleaning data, but regular expression evaluation is fast and easy. Regular expression (re) has a method called sub that operates on the leftmost characters in a string and we use this method to replace special characters with a space to split them using .split(" "), which returns the list of words in the text for us to join later to form a sentence. Not clear? Let us see this in practice.

defclean_tweet(tweet):

```
# remove special charecters and hyperlings from the string using
regular expression
return    '    '.join(re.sub("(@[A-Za-z0-9]+)|([^0-9A-Za-z    \t])|(\
w+:\/\/\S+)", " ", tweet).split())
```

Now, by passing a noisy string we can get a clean string that is helpful in performing the analysis. For polarity evaluation, we are using a TextBlob [30] package. Let us define a function called analyze_sentiment that takes a noisy tweet and returns the polarity level of the respective clean tweet. Figures 43 and 44 show the post-processing outputs of sentimental analysis.

	Tweets	retweets	likes	date	clean_tweet	polarity	sentiment
0	Hilarious! A real #CopyCat brand that copied a...	538	1643	2020-01-28 12:09:46	Hilarious A real CopyCat brand that copied alm...	0.35	Positive
1	@TusharM19278409 We will not stop working on P...	11	182	2020-01-28 09:43:45	We will not stop working on POCO F1 updates Th...	0.00	Neutral
2	The 2 words; Xtreme & Xperience defines wh...	164	1273	2020-01-28 06:43:44	The 2 words Xtreme amp Xperience defines what ...	0.00	Neutral
3	@RaghuReddy505 Haha	2	59	2020-01-28 06:25:03	Haha	0.20	Positive
4	We're here to redefine the value of X. Get rea...	394	3550	2020-01-27 06:01:05	We re here to redefine the value of X Get read...	0.20	Positive

Fig. 43 After data processing

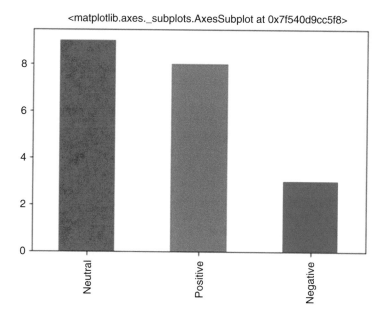

Fig. 44 Plots of bar graph for neutral, positive, and negative using tweet API

defanalyze_sentiment(tweet):

```
analysis = TextBlob(clean_tweet(tweet))
returnanalysis.sentiment.polarity
```

The polarity level ranges from −1 to 1, polarity <0 being a negative degree, polarity = 0 being a neutral degree, and polarity >1 being a positive degree.
  def degree(polarity):

```
if(polarity>0):
return 'Positive'
elif polarity==0:
return 'Neutral'
else:
return 'Negative'
```

We can include clean data, polarity, and sentiment (polarity degree) into our df Data Frame.:

```
df['clean_tweet'] = np.array([clean_tweet(tweet) for tweet in
df['Tweets']])
df['polarity'] = np.array([analyze_sentiment(tweet) for tweet in
df['Tweets']])
df['sentiment'] = np.array([degree(polarity) for polarity in
df['polarity']])
df.head()
```

Now let us see how many positives, neutrals, and negatives are there in our data by simply plotting a bar graph:

```
df['sentiment'].value_counts().plot(kind='bar',color=['r','g'
,'b'])
```

Using tweepy Cursor, streamListener, and a Little OOPS Implementation
Tweepy has a streamListener class that allows us to stream live data one after another. Since the data is live and leads to large data we can [31] write them to a file, but for this time let us limit the tweets on data stream itself and save them in a list named tweets_data. Here we are going to implement a little object-oriented programming for easy reference. Initially, we need to initiate a listener class that takes tweepy. streamListener as a parameter, which will have a constructer and 3 tweepy recogniz-able methods on_data(), on_error(), on_status(). Their working is same as their same

	Tweets	retweets	likes	date
0	RT @DPP_PKB: Bersatu Menghadapi Endemik Virus ...	0	0	Wed Jan 29 09:47:57 +0000 2020
1	RT @Pierrefirdaus: 'Vaksin' yang paling kuat a...	0	0	Wed Jan 29 09:47:57 +0000 2020
2	RT @xicograziano: Cara, esse Zé de Abreu tem u...	0	0	Wed Jan 29 09:47:57 +0000 2020
3	RT @parcha_p: งานเข้าแล้วมั้ยละ เมื่อไทยถูกมอ...	0	0	Wed Jan 29 09:47:57 +0000 2020
4	Bueno queda cancelada mi experiencia este vera...	0	0	Wed Jan 29 09:47:57 +0000 2020

Fig. 45 Tweets, retweets, and likes input data

	Tweets	retweets	likes	date	clean_tweet	polarity	sentiment
0	RT @DPP_PKB: Bersatu Menghadapi Endemik Virus ...	0	0	Wed Jan 29 09:47:57 +0000 2020	RT PKB Bersatu Menghadapi Endemik Virus Corona ...	0.0	Neutral
1	RT @Pierrefirdaus: Vaksin yang paling kuat a ...	0	0	Wed Jan 29 09:47:57 +0000 2020	RT Vaksin yang paling kuat adalah doa jangan p ...	0.0	Neutral
2	RT @xicograziano: Cara, esse Zé de Abreu tem u ...	0	0	Wed Jan 29 09:47:57 +0000 2020	RT Cara esse Z de Abreu tem uma fixa o comigo ...	0.0	Neutral
3	RT @parcha_p: งานเข้าแล้วมั้ยละ เมื่อไทยถูกมอ ...	0	0	Wed Jan 29 09:47:57 +0000 2020	RT p	0.0	Neutral
4	Bueno queda cancelada mi experiencia este vera ...	0	0	Wed Jan 29 09:47:57 +0000 2020	Bueno queda cancelada mi experiencia este vera ...	0.0	Neutral

Fig. 46 Data after post processing

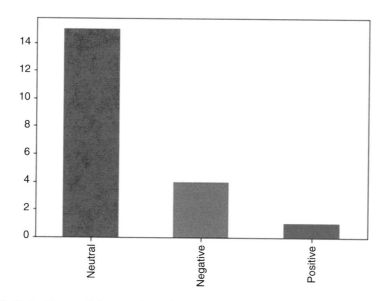

Fig. 47 Plots of bar graph for neutral, positive, and negative using tweepy cursor, streamListener

suggests. Tweepy.streamListerner will trigger these methods while the data is streaming. on_data() we will append the incoming data to tweets_data list until we reach a
limit and when on_error() and on_status() we simply print those. Figures 45, 46, and
47 provide the code implementation and output of the tweepystreamlistener.

classtwitterStreamListener(tweepy.StreamListener):

```
"""
Listener class to print received tweets
"""
def __init__(self,limit):
self.count=0

defon_data(self,data):
self.count += 1
tweets_data.append(data)
if(self.count>=limit):
print("Data Extracted:",len(tweets_data))
return False

defon_error(self,status):
print("Error: ", status)

defon_status(self, status):
print("Status: ",status.text)
```

You can add code for writing streaming data into a file under or in place of the line tweets_data.append(data).

In the main class, we have to create an instance of twitterStreamListener in order to use the resources of that class. Since we [32] are using a constructor method in that class, we need to create an object with a limit as parameter. Remember that this class is streaming the data only when tweepy allows access. So here we need to use the tweepy.Stream() module to authenticate and pass the class object where streaming methods are initiated. As mentioned earlier, all the streaming data is live containing unnecessary tweets, so we need to filter through those tweets to get the desired tweets. twitter.Stream has an inbuild method called filter() to filter tweets when given a list of desired keywords/hash tags:

```
tweets_data = []
if __name__ == "__main__":

 # TWEETS RELATED TO HASHTAG
limit = 20
hash_tag_list = ["coronavirus"]
TwitterStreamListener = twitterStreamListener(limit)
twitterStream = tweepy.Stream(auth, TwitterStreamListener)
twitterStream.filter(track=hash_tag_list)
```

tweets_data contains all the tweets, and as we have done in the previous method, we need to parse the data into a Data Frame but here the tweeter_data contains a list of dictionary strings that we need to convert to dictionary using a json package and then push them into the Data Frame:

```
importjson
df = pd.DataFrame(data = [json.loads(tweet)['text'] for tweet in
tweets_data], columns=['Tweets'])
df['retweets'] = np.array([json.loads(tweet)['retweet_count'] for
tweet in tweets_data])
df['likes']  =  np.array([json.loads(tweet)['favorite_count']  for
tweet in tweets_data])
df['date'] = np.array([json.loads(tweet)['created_at'] for tweet
in tweets_data])
df.head()
```

Cleaning(), analyze_sentiment(), degree() functions and methodology is same and shown in (Fig. 46).

By plotting we get (Fig. 47),

Similarly, for getting the tweets of a particular user, we can use the tweepy. Cursor method (Figs. 48, 49, and 50), which uses pagenation and works like a crawler for Twitter. The basic usage for this method is to crawl the pages of timeline of a particular page up to a number of tweets. If the user is given as none, then the authenticated user's ID should be considered:

```
classTwitterClient():
def __init__(self):
self.twitter_client = tweepy.API(auth)
 # self.twitter_user = twitter_user

defget_twitter_client_api(self):
returnself.twitter_client

defget_user_timeline_tweets(self, num_tweets, user):
tweets=[]
for tweet in tweepy.Cursor(self.twitter_client.user_timeline, id =
user).items(num_tweets):
tweets.append(tweet)
print("Data Extracted:",len(tweets))
return tweets
```

In the main class, we need to create an instance of the TwitterClient() class and through that we can call the get_user_timeline_tweets() method by passing the number of tweets and username of a specific user; if you want your timeline, then give "none" in place of username.

	Tweets	retweets	likes	date
0	You'll be charged ⚡ up if we reveal what #POC...	231	913	2020-01-29 06:34:17
1	Hilarious! A real #CopyCat brand that copied a...	668	1968	2020-01-28 12:09:46
2	@TusharM19278409 We will not stop working on P...	11	189	2020-01-28 09:43:45
3	The 2 words; Xtreme & Xperience defines wh...	177	1375	2020-01-28 06:43:44
4	@RaghuReddy505 Haha	2	61	2020-01-28 06:25:03

Fig. 48 Data input with tweets, retweets, and likes

	Tweets	retweets	likes	date	clean_tweet	polarity	sentiment
0	You'll be charged ⚡ up if we reveal what #POC...	231	913	2020-01-29 06:34:17	You ll be charged up if we reveal what POCOX2 ...	0.00	Neutral
1	Hilarious! A real #CopyCat brand that copied a...	668	1968	2020-01-28 12:09:46	Hilarious A real CopyCat brand that copied alm...	0.35	Positive
2	@TusharM19278409 We will not stop working on P...	11	189	2020-01-28 09:43:45	We will not stop working on POCO F1 updates Th...	0.00	Neutral
3	The 2 words; Xtreme & Xperience defines wh...	177	1375	2020-01-28 06:43:44	The 2 words Xtreme amp Xperience defines what ...	0.00	Neutral
4	@RaghuReddy505 Haha	2	61	2020-01-28 06:25:03	Haha	0.20	Positive

Fig. 49 Data after processing

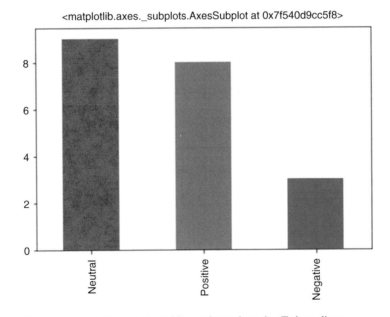

Fig. 50 Plots of bar graph for neutral, positive, and negative using Twitter client

```
tweets_data = []
if __name__ == "__main__":

    # TWEETS RELATED TO A PERTICULAT USER
    twitter_client = TwitterClient()
    tweets = twitter_client.get_user_timeline_tweets(20,'cmanmohan')
```

Tweets variable consists of data in the form of a list containing all [33] the tweets streamed. Like we have done in the previous stages, we need to convert the streamed data into a pandas Data Frame:

```
df = pd.DataFrame(data = [tweet.text for tweet in tweets],
columns=['Tweets'])
df['retweets'] = np.array([tweet.retweet_count for tweet in
tweets])
df['likes'] = np.array([tweet.favorite_count for tweet in tweets])
df['date'] = np.array([tweet.created_at for tweet in tweets])
df.head()
```

Cleaning(), analyze_sentiment(), degree() functions and methodology is same and shown in (Fig. 49).

By plotting we get (Fig. 50),

References

1. Aizerman, M.A., Braverman, E.M. and Rozoner, L.I. Theoretical foundations of the potential function method in pattern recognition learning. Automation and Remote Control, 25:821–837, 1964
2. Baron, R.A., & Ensley, M.D. 2005. Opportunity recognition as the detection of meaningful patterns: Evidence from the prototypes of novice and experienced entrepreneurs. Manuscript under review
3. Deep Learning for Computer Vision: Expert techniques to train advanced neural networks using TensorFlow and Keras. [Authors: RajalingappaaShanmugamani]
4. Deep Learning in Python: Master Data Science and Machine Learning with Modern Neural Networks written in Python, Theano, and TensorFlow. [Authors: LazyProgrammer]
5. Deep learning quick reference : useful hacks for training and optimizing deep neural networks with TensorFlow and Keras. [Authors: Bernico, Mike]
6. Deep Learning with Applications Using Python: Chatbots and Face, Object, and Speech Recognition with Tensorflow and Keras. [Authors: Navin Kumar Manaswi]
7. Deep Learning with TensorFlow: Explore neural networks with Python [Authors: Giancarlo Zaccone, Md. RezaulKarim, Ahmed Menshawy]
8. Devroye, L., Gyorfi, L. and Lugosi, G. A Probabilistic Theory of Pattern Recognition. Springer Verlag, Applications of Mathematics Vol. 31, 1996.
9. Hands-On Deep Learning for Images with TensorFlow: Build intelligent computer vision applications using TensorFlow and Keras [Authors: Will Ballard]
10. Hands-On Machine Learning with Scikit-Learn and TensorFlow: Concepts, Tools, and Techniques to Build Intelligent Systems. [Author: AurélienGéron]
11. Hands-On Transfer Learning with Python Implement Advanced Deep Learning and Neural Network Models Using TensorFlow and Keras [Authors: DipanjanSarkar, Raghav Bali, TamoghnaGhosh]
12. Hands-on unsupervised learning with Python : implement machine learning and deep learning models using Scikit-Learn, TensorFlow, and more [Authors: Bonaccorso, Giuseppe]
13. Intelligent mobile projects with TensorFlow : build 10+ artificial intelligence apps using TensorFlow Mobile and Lite for iOS, Android, and Raspberry Pi. [Authors: Tang, Jeff]
14. Intelligent Projects Using Python: 9 real-world AI projects leveraging machine learning and deep learning with TensorFlow and Keras. [Authors: SantanuPattanayak]

15. Internet of Things for Industry 4.0, EAI, Springer, Editors, G. R. Kanagachidambaresan, R. Anand, E. Balasubramanian and V. Mahima, Springer.
16. Krestinskaya O, Bakambekova A, James AP (2019) Amsnet: analog memristive system architecture for mean-pooling with dropout convolutional neural network. In: IEEE international-conference on artificial intelligence circuits and systems
17. Krestinskaya O, Salama KN, James AP (2018) Learning in memristive neural network architectures using analog backpropagation circuits. IEEE Trans Circuits Syst I: Regul Pap 1–14.https://doi.org/10.1109/TCSI.2018.2866510
18. Learn TensorFlow 2.0: Implement Machine Learning And Deep Learning Models With Python. [Authors: Pramod Singh, Avinash Manure]
19. Li Y, Wang Z, Midya R, Xia Q, Yang JJ (2018) Review of memristor devices in neuromorphic computing: materials sciences and device challenges. J Phys D: ApplPhys 51(50):503002
20. Liao Q, Poggio T (2016) Bridging the gaps between residual learning, recurrent neural networks and visual cortex. arXiv:1604.03640
21. Lippmann R (1987) An introduction to computing with neural nets. IEEE ASSP Mag 4(2):4–22
22. Ma J, Tang J (2017) A review for dynamics in neuron and neuronal network. Nonlinear Dyn 89(3):1569–1578
23. Maan AK, Jayadevi DA, James AP (2017) A survey of memristive threshold logic circuits. IEEE Trans Neural Netw Learn Syst 28(8):1734–1746
24. Mastering TensorFlow 1.x: Advanced machine learning and deep learning concepts using TensorFlow 1.x and Keras. [Author: Armando Fandango]
25. McCulloch WS, Pitts W (1943) A logical calculus of the ideas immanent in nervous activity. Bull Math Biophys 5(4):115–133
26. Osuna, E. and Girosi. F. Reducing the run-time complexity of support vector machines. In International Conference on Pattern Recognition (submitted), 1998
27. Osuna, E., Freund, R. and Girosi, F. Training support vector machines: an application to face detection. In IEEE Conference on Computer Vision and Pattern Recognition, pages 130 – 136, 1997.
28. Practical Computer Vision Applications Using Deep Learning with CNNs: With Detailed Examples in Python Using TensorFlow and Kivy. [Author: Ahmed Fawzy Gad]
29. Practical Deep Learning for Cloud, Mobile, and Edge: Real-World AI & Computer-Vision Projects Using Python, Keras&TensorFlow [Authors: AnirudhKoul, Siddha Ganju, MeherKasam]
30. Python Deep Learning: Exploring deep learning techniques, neural network architectures and GANs with PyTorch, Keras and TensorFlow. [Authors: Ivan Vasilev, Daniel Slater, GianmarioSpacagna, Peter Roelants, Valentino Zocca]
31. Ren S, He K, Girshick RB, Sun J (2017) Faster r-cnn: towards real-time object detection with region proposal networks. IEEE Trans Pattern Anal Mach Intell 39(6):1137–1149
32. Smola, A. and Sch¨olkopf, B. On a kernel-based method for pattern recognition, regression, approximation and operator inversion. Algorithmica (to appear), 1998.
33. TensorFlow 1.x Deep Learning Cookbook: Over 90 unique recipes to solve artificial-intelligence driven problems with Python. [Authors: Antonio Gulli, AmitaKapoor]

Programming Tensor Flow with Single Board Computers

G. R. Kanagachidambaresan, Kolla Bhanu Prakash, and V. Mahima

1 Introduction

More SBC has come in today's market for rapid prototyping and modelling ad hoc solutions. Table 1 illustrates the recent most sold SBCs.

Single board computers are mainly used for limited applications and mainly sophisticated IoT and edge applications [1, 4, 5, 6]. The single computer has very low capability and with resources starving in nature. The single board computers are enabled with general purpose input and output (GPIO) pins to handle electrical signals and communicate with a wire and wirelessly with nearby cyber physical systems (CPS) [7, 8, 9, 10, 11]. Figure 1 illustrates the NVIDIA Jetson Nano SBC with a 40-pin GPIO. The SBC has the UBUNTU operating system with CUDA framework to make GPU computation [12, 13, 14, 15]. The Jetson Nano is enabled with a CUDA framework facility to make faster computation.

The package for GPIO activation is Jetson.GPIO, and the following program illustrates how to switch on and off an LED using GPIO of Jetson Nano:

G. R. Kanagachidambaresan (✉) · V. Mahima
Vel Tech Rangarajan Dr Sagunthala R&D Institute of Science and Technology,
Chennai, Tamil Nadu, India

K. B. Prakash
KL Deemed to be University, Vijayawada, AP, India

© Springer Nature Switzerland AG 2021
K. B. Prakash, G. R. Kanagachidambaresan (eds.), *Programming with TensorFlow*, EAI/Springer Innovations in Communication and Computing,
https://doi.org/10.1007/978-3-030-57077-4_12

Table 1 Recent SBCs comparison table

S. No	Name of SBC	Processor	Speed	RAM	Operating system
1	Raspberry pi 4	Broadcom BCM2711	1.5 GHz	2GB	Raspian
2	Ordroid	Samsung Exynos5422 cortex™-A15	2 GHz	2GB	Linux and android
3	NVIDIA Jetsonnano	Quad-core ARM® cortex®-A57 MPCore processor	1.6GHz	4GB	Ubuntu
4	NVIDIA Xavier	8-core ARM v8.2 64-bit CPU	2.26 GHz	12 GB	Ubuntu

Fig. 1 NVIDIA Jetson nano single board computer (SBC)

```
importJetson.GPIO as GPIO
//Package Importing for handling GPIO pins of the SBC
import time
//for delay time setting
GPIO.setmode(GPIO.BOARD)
//mapping the pins as per board numbering
bulp = [18]
//pin number 18 assigned to glow bulb
GPIO.setup(bulp, GPIO.OUT, initial=GPIO.LOW)
//Initializing pin 18 as output pin with zero initial condition,
so that bulp will be in off condition when it starts
GPIO.output(bulp, GPIO.HIGH)
//pin 18 is assigned high
time.sleep(10);
```

```
//make the pin in high state for ten seconds
GPIO.output(bulp, GPIO.LOW)
// bulp off.
```

The tensor flow installation in GPUs can be done through the following command:

```
pip3 install tensorflow-gpu #for GPU installation in NVIDIA
```

Figure 2 illustrates the NVIDIA Xavier single board computer, more capable with high resources [16] when compared with the Jetson Nano board. The XAVIER board is mainly used for image processing and automatic driving assistance purposes to understand and recognize the images and to acknowledge the system accordingly.

Both the NVIDIA Jetson and Xavier modules are [17] enabled with camera interface options. The camera can be connected via i2c and USB ports. The following procedure describes the camera installation and snapshot code for taking picture from a webcam in NVIDIA SBCs.

Installation from git repository using the command.

```
git clone https://github.com/NVIDIA-AI-IOT/jetcam
cdjetcam
sudo python3 setup.py install
```

Importing python packages:

Fig. 2 NVIDIA Xavier single board computer

```
fromjetcam.csi_camera import CSICamera
cam1 = CSICamera(width=240, height=320, capture_width=1080, cap-
ture_height=720, capture_fps=20)
```

Here the image display height is specified as 329 and the width is specified as 240. The width of capturing is 1080 and the height of capturing is 720. The variable cam1 is mentioned as CSI for ribbon cable camera.

```
fromjetcam.usb_camera import USBCamera
cam1 = USBCamera(capture_device=1)
```

The variable cam1 is set for a USB camera. The code used to open the camera shutter is given below:

```
image = camera.read()
```

2 CUDA Programming in NVIDIA

Both the NVIDIA SBCs are capable of working on the CUDA framework. The nvcc compiler essential for performing the CUDA [18] operation is to be exported. The operating system already comes with CUDA and the following command can be used to export the path of language-CUDA:

```
export PATH=${PATH}:/usr/local/cuda/bin
export LD_LIBRARY_PATH=${LD_LIBRARY_PATH}:/usr/local/cuda/lib64
```

The verification of the above is done through the following command in terminal box:

```
nvcc -v
```

Here in the programming, the sentence to be executed in the GPU should be mentioned with __global__ syntax identifier. The TensorRT package can accelerate the performance of GPU and provide lower computation time in matrix calculations. The following code describes the simple matrix multiplication code in the CUDA environment in NVIDIA jetsonnano SBC:

```
@cuda.jit
""" initializing CUDA
defmatrixmul(mat1, mat2, matresult):
  """square matrix multiplication (mat1* mat2 = matresult)
p, q = cuda.grid(2)
if p <matresult.shape[0] and q <matresult.shape[1]:
temp = 0.
for k in range(mat1.shape[1]):
temp += mat1[p, k] * mat2[k, q]
matresult[i, j] = temp
```

The SBC has various industrial needs and is presently solving various industrial problems. The following section discusses some of the related industrial problems

3   Prepackaging Inspection Module for Industry 4.0

This below module aims on developing a prepackaging inspection machine for packaging goods before being dispatched. This is a low cost cyber physical system which verifies the content in the package and performs a check before wrapping for shipment. The kit is trained with the images of toolset in Raspberry pi4 and Jetsonnano using deep learning, tensor flow, keras, open cv, and GPIO. The images for the inspection are captured by a camera placed on the top center of the computing machine. The images captured are being trained by the precaptured and stored images. The captured images undergo a series of training phases to check and verify whether the toolsets are perfectly aligned and placed. In the first step of the process, the toolset will be placed in the inspection machine to capture it. Then the captured images are segmented into two types: (1) right alignment and (2) wrong alignment. The correct images are the ones which are perfectly aligned and placed in the perfect order. The wrong images are the ones which are misplaced, misaligned, and in some cases certain tools from the toolset are missed during the prepackaging. The results of the correct and wrong model are indicated by the machine with the help of green and red LEDs. If the image is correct, it will blink green else it will blink red. The region of interest for the camera is 32x24 cm.

Figure 3 illustrates the hardware design of the module. The module is trained with tensor flow and deep learning modules.

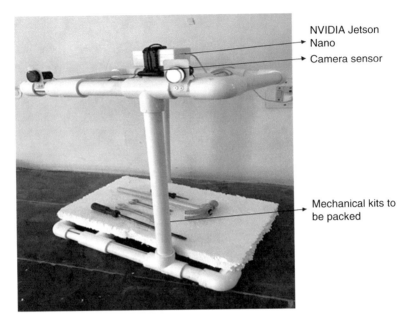

Fig. 3 Tensor flow–based pre-inspection machine

Figure 4 is the flowchart proposed for the prepackaging inspection machine. The equipment is moved inside the region of interest through conveyer mechanism. The image is first given as input through a camera sensor properly mounted covering the region of interest, then the object is recognized and classified through tensor flow deep learning for better accuracy. After classification, the image is checked for correct alignment and wrong alignment through deep learning and it helps in taking decision. If the image is correct, it permits and indicates a green light and rotates the conveyer in the forward direction, if the image is wrong it does not permit and indicates a red light. Then the number of correct images and wrong images detected are saved and end the process. The counter is integrated with the Google cloud service and it is able to indicate the production rate to the subscribers. The system is highly accurate and presently trained with more than 3000 images of a single product kit set. The same module can be easily trained for other similar applications on packaging.

An input image is given and it is stored in .jpeg file format. Then they are tested and trained. After testing and training the records are saved. With the records the images are trained and evaluate the new input images. Figures 5 and 6 elucidate that when the camera senses the correct aligned tool kit, it gives a green signal, else it gives a red signal.

Figure 7 illustrates the verification time taken by each single board computer. NVIDIA Xavier has the least computational time and checks more packages per minute.

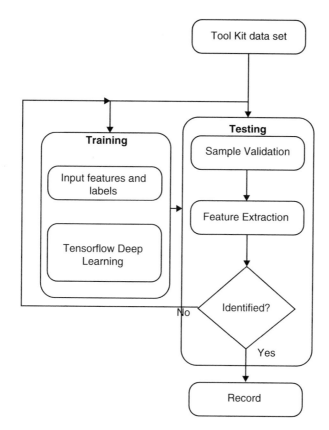

Fig. 4 Flowchart diagram

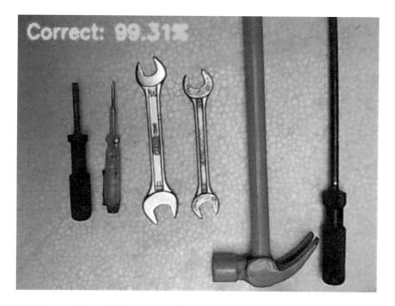

Fig. 5 Real-time results verified as correct alignment

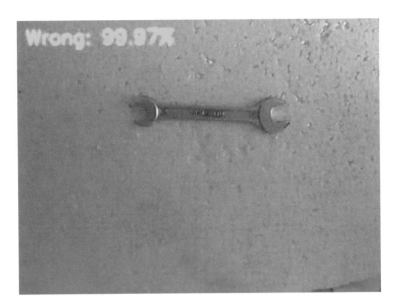

Fig. 6 Real-time results verified as wrong alignment

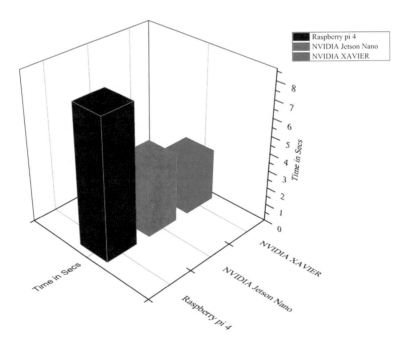

Fig. 7 Time complexity of different SBCs

4   Fish Geopositioning System for Industry 4.0

This module aims in developing automatic fish positioning system for fish packaging industry, our motive of this invention are to reduce the difficulties faced by the labor during cutting process, fastening of packaging work, and to protect the labor from eye sight stress, it causes like myopia, glaucoma, hyperopia and smell allergies with the help of the technology of neural network.

The first step for packaging is to catch the fish with huge quantity for processing, once the fishes are catched, then fishes are taken to the fish processing machine and passed into the conveyer and the fishes will start to move over the conveyer track belt. The fish are cut and cleaned before preservation and transport. The randomness of fish in the conveyer makes the cutting machine more human and intervention is required to rotate and properly position the fish for the cutting process.

The position detection of fish is detected by camera, using the convolution neural network; if the position is wrong then it is corrected with the help of a motor so that the described problem faced by labor will be solved, the efficiency of work will be increased, and the task completed faster. Figure 8 illustrates the working process of the automatic fish positioning system using deep convolutional neural network and IoT.

Figure 9 illustrates the correct and wrong position of fish to be trained for the robot. The convolution neural network understands the current position of the fish and makes N rotation with a stepper motor drive to rotate the fish to the correct position.

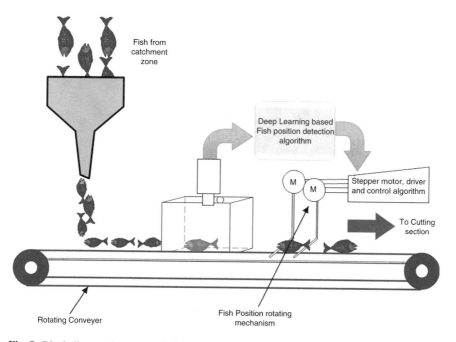

Fig. 8 Block diagram for automatic fish positioning system

Fig. 9 Data set for training the fish position

posi+_pos: 99.59%

Fig. 10 Real-time results verified – correct position of fish

Figure 10 elucidates the results from the SBC, and the SBC is directly connected with servo motor through GPIO and commands the servo to make certain necessary rotations for cutting and cleaning purposes. Figure 11 shows the wrong position of the fish. Figure 12 illustrates the training loss with epoch for the modeled CNN.

Figure 13 illustrates the time taken by a single board computer to complete single verification. NVIDIA Xavier takes minimum time to verify the given input images.

Fig. 11 Wrong position
identification of fish

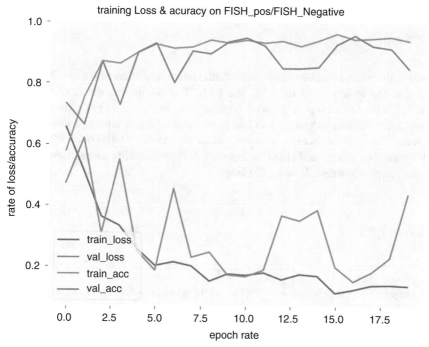

Fig. 12 Training loss and accuracy on fish position

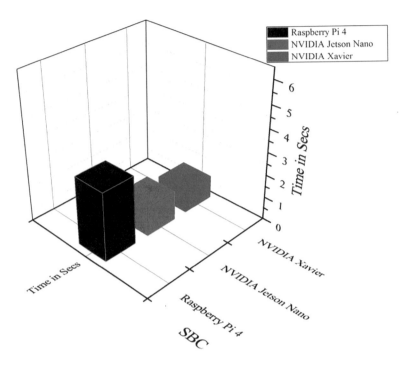

Fig. 13 Verification time for each SBC

5 Conclusion

The single board computers have become the necessary element for rapid prototyping and are widely used for edge analytics. The computing devices in SoC are enabled with fast computing and a communication device in par with the client. Apart from all the constraints, it is low power consuming in nature. The application of tensor flow in SBC is enormous and can serve smart city applications. The industry Internet of Things and standard Industry 4.0 provide a bigger platform for SBC with opencv and tensor flow applications.

References

1. Deep Learning Pipeline: Building A Deep Learning Model With TensorFlow [Authors: Hisham El-Amir, Mahmoud Hamdy]
2. Deep Learning with TensorFlow: Explore neural networks with Python [Authors: Giancarlo Zaccone, Md. RezaulKarim, Ahmed Menshawy]
3. Goodfellow I, Bengio Y, Courville A, Bengio Y (2016) Deep learning, vol 1. MIT Press, Cambridge

4. Grandison T, Sloman M (2000) A survey of trust in internet applications. IEEE CommunSurv Tutor 3(4):2–16

5. Hands-On Deep Learning for Images with TensorFlow: Build intelligent computer vision applications using TensorFlow and Keras [Authors: Will Ballard]

6. Hands-On Machine Learning with Scikit-Learn and TensorFlow: Concepts, Tools, and Techniques to Build Intelligent Systems. [Author: AurélienGéron]

7. Hands-On Transfer Learning with Python Implement Advanced Deep Learning and Neural Network Models Using TensorFlow and Keras [Authors: DipanjanSarkar, Raghav Bali, TamoghnaGhosh]

8. Hands-on unsupervised learning with Python: implement machine learning and deep learning models using Scikit-Learn, TensorFlow, and more [Authors: Bonaccorso, Giuseppe]

9. Internet of Things for Industry 4.0, EAI, Springer, Editors, G. R. Kanagachidambaresan, R. Anand, E. Balasubramanian and V. Mahima, Springer.

10. Learn TensorFlow 2.0: Implement Machine Learning And Deep Learning Models With Python. [Authors: Pramod Singh, Avinash Manure]

11. Learning TensorFlow [Authors: Tom Hope, Yehezkel S. Resheff&Itay Lieder]

12. Mastering TensorFlow 1.x: Advanced machine learning and deep learning concepts using TensorFlow 1.x and Keras. [Author: Armando Fandango]

13. Practical Computer Vision Applications Using Deep Learning with CNNs: With Detailed Examples in Python Using TensorFlow and Kivy. [Author: Ahmed Fawzy Gad]

14. Practical Deep Learning for Cloud, Mobile, and Edge: Real-World AI & Computer-Vision Projects Using Python, Keras&TensorFlow [Authors: AnirudhKoul, Siddha Ganju, MeherKasam]

15. Python Deep Learning: Exploring deep learning techniques, neural network architectures and GANs with PyTorch, Keras and TensorFlow. [Authors: Ivan Vasilev, Daniel Slater, GianmarioSpacagna, Peter Roelants, Valentino Zocca]

16. Python Machine Learning: Machine Learning and Deep Learning with Python, scikit-learn, and TensorFlow [Authors: Sebastian Raschka, VahidMirjalili]

17. TensorFlow 1.x Deep Learning Cookbook: Over 90 unique recipes to solve artificial-intelligence driven problems with Python. [Authors: Antonio Gulli, AmitaKapoor]

18. TensorFlow for Machine Intelligence_ A Hands-On Introduction to Learning Algorithms [Authors: Sam Abrahams, DanijarHafner, Erik Erwitt, Ariel Scarpinelli]

Appendices

Appendix 1

1. McCulloch-Pitts Neuron Model

```python
import numpy as np
import matplotlib.pyplot as plt

w=[1,1]
ba,bo=-1.5,-0.5
def And(inp):
    print(len(inp))
    l=[]
    for i in range(len(inp)):
        y=inp[i][0]*w[0]+inp[i][1]*w[1]+ba
        if(y>=0):
            l.append(1)
        else:
            l.append(0)
    return l
```

© Springer Nature Switzerland AG 2021
K. B. Prakash, G. R. Kanagachidambaresan (eds.), *Programming with
TensorFlow*, EAI/Springer Innovations in Communication and Computing,
https://doi.org/10.1007/978-3-030-57077-4

```python
def Or(inp):
    o=[]
    for i in range(len(inp)):
        y=inp[i][0]*w[0]+inp[i][1]*w[1]+bo
        if(y>=0):
            o.append(1)
        else:
            o.append(0)
    return o
```

```python
inp=np.array([[0,0],[0,1],[1,0],[1,1]])
a=And(inp)
b=Or(inp)
print(a)
print(b)
```

```
4
[0, 0, 0, 1]
[0, 1, 1, 1]
```

```python
x,y=zip(*inp)
```

```python
import matplotlib.lines as lines
plt.scatter(x,y)
plt.plot([1,0],[0,1],linestyle='solid')
```

```
[<matplotlib.lines.Line2D at 0x1f70e6205c0>]
```

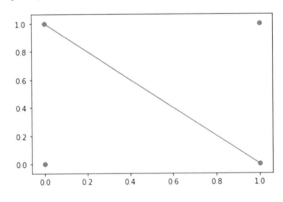

2. Multi-Layer Perceptron, Keras Intuition, Backpropagation, Activation function, and Gradient Descent

```python
import pandas as pd
```

```python
ds=pd.read_csv("fruit_name_color.csv")
```

```python
f=["mass","width","height","color_score"]
x=ds[f]
y=ds["fruit_label"]
```

```python
from sklearn.model_selection import train_test_split
```

```python
x_train,x_test,y_train,y_test=train_test_split(x,y)
```

```python
from sklearn import linear_model as lm
```

```python
model=lm.LinearRegression()
```

```python
model.fit(x_train,y_train)
LinearRegression(copy_X=True, fit_intercept=True, n_jobs=None, normalize=False)
```

```python
p=model.predict(x_test)
```

3. (a). Performing Classification and Regression Using Artificial Neural Network

```python
import pandas as pd
import numpy as np
```

```python
ds=pd.read_csv("Churn_Modelling.csv")
#ds
```

```python
x=ds.iloc[:,4:13].values
```

```python
x
```

```python
array([['France', 'Female', 42, ..., 1, 1, 101348.88],
       ['Spain', 'Female', 41, ..., 0, 1, 112542.58],
       ['France', 'Female', 42, ..., 1, 0, 113931.57],
       ...,
       ['France', 'Female', 36, ..., 0, 1, 42085.58],
       ['Germany', 'Male', 42, ..., 1, 0, 92888.52],
       ['France', 'Female', 28, ..., 1, 0, 38190.78]], dtype=object)
```

```
y=ds.iloc[:,13:].values
```

```
y
```

```
array([[1],
       [0],
       [1],
       ...,
       [1],
       [1],
       [0]], dtype=int64)
```

```
from sklearn import preprocessing
```

```
le = preprocessing.LabelEncoder()
```

```
x[:,0]=le.fit_transform(x[:,0])
```

```
x[:,1]=le.fit_transform(x[:,1])
```

```
x
```

```
array([[0, 0, 42, ..., 1, 1, 101348.88],
       [2, 0, 41, ..., 0, 1, 112542.58],
       [0, 0, 42, ..., 1, 0, 113931.57],
       ...,
       [0, 0, 36, ..., 0, 1, 42085.58],
       [1, 1, 42, ..., 1, 0, 92888.52],
       [0, 0, 28, ..., 1, 0, 38190.78]], dtype=object)
```

```
from sklearn.preprocessing import OneHotEncoder
oh=OneHotEncoder(categorical_features=[0])#represents which column to be onehot encoded
x=oh.fit_transform(x).toarray()
```

```
x
```

```
array([[1.0000000e+00, 0.0000000e+00, 0.0000000e+00, ..., 1.0000000e+00,
        1.0000000e+00, 1.0134888e+05],
       [0.0000000e+00, 0.0000000e+00, 1.0000000e+00, ..., 0.0000000e+00,
        1.0000000e+00, 1.1254258e+05],
       [1.0000000e+00, 0.0000000e+00, 0.0000000e+00, ..., 1.0000000e+00,
        0.0000000e+00, 1.1393157e+05],
       ...,
       [1.0000000e+00, 0.0000000e+00, 0.0000000e+00, ..., 0.0000000e+00,
        1.0000000e+00, 4.2085580e+04],
       [0.0000000e+00, 1.0000000e+00, 0.0000000e+00, ..., 1.0000000e+00,
        0.0000000e+00, 9.2888520e+04],
       [1.0000000e+00, 0.0000000e+00, 0.0000000e+00, ..., 1.0000000e+00,
        0.0000000e+00, 3.8190780e+04]])
```

```python
from sklearn.model_selection import train_test_split
x_train,x_test,y_train,y_test=train_test_split(x,y,test_size=0.20
,random_state=0)
```

```python
#Dependencies
import keras
from keras.models import Sequential
from keras.layers import Dense
# Neural network
model = Sequential()
model.add(Dense(11, input_dim=11, activation='relu'))
model.add(Dense(5, activation='relu'))
model.add(Dense(1, activation='sigmoid'))
```

```python
model.compile(loss='binary_crossentropy', optimizer='adam', metrics=['accuracy'])
y_pred = model.predict(x_test)
pred = list()
for i in range(len(y_pred)):
    pred.append(np.argmax(y_pred[i]))
test = list()
for i in range(len(y_test)):
    test.append(np.argmax(y_test[i]))
from sklearn.metrics import accuracy_score
a = accuracy_score(pred,test)
print('Accuracy is:', a*100)
```

```
Accuracy is: 100.0
```

(b). Performing Classification and Regression Using Artificial Neural Network

```python
from pandas import read_csv
from keras.models import Sequential
from keras.layers import Dense
from keras.wrappers.scikit_learn import KerasRegressor
from sklearn.model_selection import cross_val_score
from sklearn.model_selection import KFold
from sklearn.preprocessing import StandardScaler
from sklearn.pipeline import Pipeline
# load dataset
dataframe = read_csv("houseprice.csv",header=None)
dataset = dataframe.values
# split into input (X) and output (Y) variables
X = dataset[:,0:12]
Y = dataset[:,12]

# define base model
def baseline_model():
    # create model
    model = Sequential()
    model.add(Dense(12, input_dim=12, kernel_initializer='normal', activation='relu'))
    model.add(Dense(1, kernel_initializer='normal'))
    # Compile model
    model.compile(loss='mean_squared_error', optimizer='sgd')
    return model
# evaluate model with standardized dataset
estimators = []
estimators.append(('standardize', StandardScaler()))
estimators.append(('mlp', KerasRegressor(build_fn=baseline_model, epochs=50,
                                         batch_size=5, verbose=0)))

pipeline = Pipeline(estimators)
kfold = KFold(n_splits=10)
results = cross_val_score(pipeline, X, Y, cv=kfold)
print("Standardized: %.2f (%.2f) MSE" % (results.mean(), results.std()))
```

Standardized: -135.02 (311.17) MSE

4. (a). Principal Component Analysis

```python
# importing required libraries
import numpy as np
import matplotlib.pyplot as plt
import pandas as pd
```

```python
# importing or loading the dataset
dataset = pd.read_csv('wine.csv')
```

```python
# distributing the dataset into two components X and Y
X = dataset.iloc[:, 0:13].values
y = dataset.iloc[:, 13].values
```

```python
# Splitting the X and Y into the
# Training set and Testing set
from sklearn.model_selection import train_test_split
X_train, X_test, y_train, y_test = train_test_split(X, y, test_size = 0.2, random_state = 0)
```

```python
# performing preprocessing part
from sklearn.preprocessing import StandardScaler
sc = StandardScaler()

X_train = sc.fit_transform(X_train)
X_test = sc.transform(X_test)
```

```python
# Applying PCA function on training
# and testing set of X component
from sklearn.decomposition import PCA

pca = PCA(n_components = 2)

X_train = pca.fit_transform(X_train)
X_test = pca.transform(X_test)

explained_variance = pca.explained_variance_ratio_
```

```python
# Fitting Logistic Regression To the training set
from sklearn.linear_model import LogisticRegression

classifier = LogisticRegression(random_state = 0)
classifier.fit(X_train, y_train)
```

```
Out[6]: LogisticRegression(C=1.0, class_weight=None, dual=False, fit_intercept=True,
                            intercept_scaling=1, l1_ratio=None, max_iter=100,
                            multi_class='warn', n_jobs=None, penalty='l2',
                            random_state=0, solver='warn', tol=0.0001, verbose=0,
                            warm_start=False)
```

```
# Predicting the test set result using
# predict function under LogisticRegression
y_pred = classifier.predict(X_test)
y_pred
```

```
array([1285,  660, 1035, 1285,  680, 1285, 1285,  520,  562,  520,  520,
        520, 1285, 1035,  660,  562, 1285, 1065,  520, 1285,  562, 1035,
       1285,  562,  520,  562,  680,  562,  520,  560, 1285, 1285,  520,
       1285, 1285, 1035], dtype=int64)
```

```
# making confusion matrix between
# test set of Y and predicted value.
from sklearn.metrics import confusion_matrix

cm = confusion_matrix(y_test, y_pred)
cm
```

```
array([[0, 0, 0, ..., 0, 0, 0],
       [0, 0, 0, ..., 0, 0, 0],
       [0, 0, 0, ..., 0, 0, 0],
       ...,
       [0, 0, 0, ..., 0, 0, 0],
       [0, 0, 0, ..., 0, 0, 0],
       [0, 0, 0, ..., 0, 0, 0]], dtype=int64)
```

```
# Predicting the training set
# result through scatter plot
from matplotlib.colors import ListedColormap

X_set, y_set = X_train, y_train
X1, X2 = np.meshgrid(np.arange(start = X_set[:, 0].min() - 1,stop = X_set[:, 0].max() + 1,
                               step = 0.01), np.arange(start = X_set[:, 1].min() - 1,
                                                       stop = X_set[:, 1].max() + 1,
                                                       step = 0.01))
plt.contourf(X1, X2, classifier.predict(np.array([X1.ravel(), X2.ravel()]).T).reshape(X1.shape),
             alpha = 0.75, cmap = ListedColormap(('yellow', 'white', 'aquamarine')))
plt.xlim(X1.min(), X1.max())
plt.ylim(X2.min(), X2.max())
```

```
# Predicting the training set
# result through scatter plot
from matplotlib.colors import ListedColormap

X_set, y_set = X_train, y_train
X1, X2 = np.meshgrid(np.arange(start = X_set[:, 0].min() - 1,stop = X_set[:, 0].max() + 1,
                    step = 0.01), np.arange(start = X_set[:, 1].min() - 1,
                                            stop = X_set[:, 1].max() + 1,
                                            step = 0.01))
plt.contourf(X1, X2, classifier.predict(np.array([X1.ravel(), X2.ravel()]).T).reshape(X1.shape),
            alpha = 0.75, cmap = ListedColormap(('yellow', 'white', 'aquamarine')))
plt.xlim(X1.min(), X1.max())
plt.ylim(X2.min(), X2.max())
```

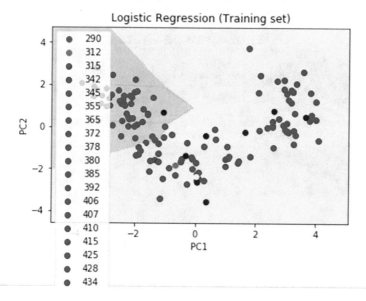

```
# Visualising the Test set results through scatter plot
from matplotlib.colors import ListedColormap

X_set, y_set = X_test, y_test

X1, X2 = np.meshgrid(np.arange(start = X_set[:, 0].min() - 1,
                    stop = X_set[:, 0].max() + 1, step = 0.01),
                np.arange(start = X_set[:, 1].min() - 1, stop = X_set[:, 1].max() + 1,
                        step = 0.01))

plt.contourf(X1, X2, classifier.predict(np.array([X1.ravel(), X2.ravel()]).T).reshape(X1.shape),
            alpha = 0.75, cmap = ListedColormap(('yellow', 'white', 'aquamarine')))

plt.xlim(X1.min(), X1.max())
plt.ylim(X2.min(), X2.max())
```

```
plt.xlim(X1.min(), X1.max())
plt.ylim(X2.min(), X2.max())

for i, j in enumerate(np.unique(y_set)):
    plt.scatter(X_set[y_set == j, 0], X_set[y_set == j, 1],
                c = ListedColormap(('red', 'green', 'blue'))(i), label = j)
# title for scatter plot
plt.title('Logistic Regression (Test set)')
plt.xlabel('PC1') # for Xlabel
plt.ylabel('PC2') # for Ylabel
plt.legend()
# show scatter plot
plt.show()
```

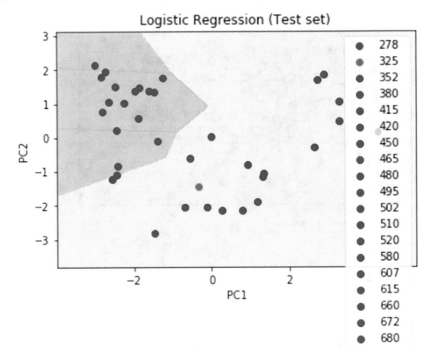

(b). Batch Normalization

```python
# importing required libraries
import numpy as np
import matplotlib.pyplot as plt
import pandas as pd
```

```python
# importing or loading the dataset
dataset = pd.read_csv('wine.csv')
# distributing the dataset into two components X and Y
X = dataset.iloc[:, 0:13].values
y = dataset.iloc[:, 13].values
```

```python
from sklearn.preprocessing import Normalizer
scaler = Normalizer().fit(X)
normalizedX = scaler.transform(X)
# summarize transformed data
np.set_printoptions(precision=3)
print(normalizedX[0::,:])
```

```
[[0.008 0.11  0.013 ... 0.044 0.008 0.03 ]
 [0.01  0.13  0.017 ... 0.043 0.01  0.033]
 [0.01  0.127 0.023 ... 0.055 0.01  0.03 ]
 ...
 [0.024 0.108 0.035 ... 0.083 0.005 0.013]
 [0.024 0.107 0.021 ... 0.076 0.005 0.013]
 [0.03  0.14  0.041 ... 0.091 0.006 0.016]]
```

```python
from sklearn.model_selection import train_test_split
X_train,X_test,y_train,y_test = train_test_split(X,y,test_size = 0.1)
```

```python
#Dependencies
import keras
from keras.models import Sequential
from keras.layers import Dense
# Neural network
model = Sequential()
model.add(Dense(16, input_dim=13, activation='relu'))
model.add(Dense(5, activation='relu'))
model.add(Dense(1, activation='sigmoid'))
```

```
y_pred = model.predict(X_test)
pred = list()
for i in range(len(y_pred)):
    pred.append(np.argmax(y_pred[i]))
test = list()
for i in range(len(y_test)):
    test.append(np.argmax(y_test[i]))
```

```
from sklearn.metrics import accuracy_score
a = accuracy_score(pred,test)
print('Accuracy is:', a*100)
```

```
Accuracy is: 100.0
```

Appendix 2

Project 1

Create a model that predicts whether or not a loan will be default using the historical data:

Problem Statement – Source - Simplilearn

Steps to Perform

Perform exploratory data analysis and feature engineering and then apply feature engineering. Follow up with a deep learning model to predict whether or not the loan will be default using the historical data.

Tasks

1. Feature Transformation
   Transform categorical values into numerical values (discrete).

2. Exploratory data analysis of different factors of the dataset.
3. Additional Feature Engineering

   You will check the correlation between features and will drop those features that have a strong correlation.

   This will help reduce the number of features and will leave you with the most relevant features.

4. Modeling

After applying Exploratory Data Analysis (EDA) and feature engineering, you are now ready to build the predictive models.

In this part, you will create a deep learning model using Keras with tensorflow backend.

```
import numpy as np
import pandas as pd

import tensorflow as tf
%matplotlib inline
```

1.14.0

```
loan_data.head()
```

	credit.policy	purpose	int.rate	installment	log.annual.inc	dti	fico	days.with.cr.line
0	1	debt_consolidation	0.1189	829.10	11.350407	19.48	737	5639.958333
1	1	credit_card	0.1071	228.22	11.082143	14.29	707	2760.000000
2	1	debt_consolidation	0.1357	366.86	10.373491	11.63	682	4710.000000
3	1	debt_consolidation	0.1008	162.34	11.350407	8.10	712	2699.958333
4	1	credit_card	0.1426	102.92	11.299732	14.97	667	4066.000000

```
loan_data.info()

<class 'pandas.core.frame.DataFrame'>
RangeIndex: 9578 entries, 0 to 9577
Data columns (total 14 columns):
credit.policy        9578 non-null int64
purpose              9578 non-null object
int.rate             9578 non-null float64
installment          9578 non-null float64
log.annual.inc       9578 non-null float64
dti                  9578 non-null float64
fico                 9578 non-null int64
days.with.cr.line    9578 non-null float64
revol.bal            9578 non-null int64
revol.util           9578 non-null float64
inq.last.6mths       9578 non-null int64
delinq.2yrs          9578 non-null int64
pub.rec              9578 non-null int64
not.fully.paid       9578 non-null int64
dtypes: float64(6), int64(7), object(1)
memory usage: 1.0+ MB
```

```
loan_data['purpose'].value_counts()

debt_consolidation    3957
all_other             2331
credit_card           1262
home_improvement       629
small_business         619
major_purchase         437
educational            343
Name: purpose, dtype: int64
```

```
from sklearn import preprocessing
le = preprocessing.LabelEncoder()
le.fit(loan_data['purpose'])
le.transform(loan_data['purpose'])
le.classes_

loan_data['purpose'] = le.fit_transform(loan_data['purpose'])
```

```
loan_data.head()
```

	credit.policy	purpose	int.rate	installment	log.annual.inc	dti	fico	days.with.cr.line	revol.ba
0	1	2	0.1189	829.10	11.350407	19.48	737	5639.958333	28854
1	1	1	0.1071	228.22	11.082143	14.29	707	2760.000000	33623
2	1	2	0.1357	366.86	10.373491	11.63	682	4710.000000	3511
3	1	2	0.1008	162.34	11.350407	8.10	712	2699.958333	33667
4	1	1	0.1426	102.92	11.299732	14.97	667	4066.000000	4740

```
loan_data.shape
```

```
(9578, 14)
```

Project 2
Build a Convolutional Neural Network (CNN) model that classifies the given pet images correctly into dog and cat images:

Problem Statement – Source - Simplilearn

Project Description and Scope
You are provided with the following resources that can be used as inputs for your model:

1. A collection of images of pets, that is, cats and dogs. These images are of different sizes with varied lighting conditions.
2. Code template containing the following code blocks:
   (a). Import modules (Part 1).
   (b). Set hyper parameters (Part 2).
   (c). Read image data set (Part 3).
   (d). Run tensorflow model (Part 4).

You are expected to write the code for CNN image classification model (between Parts 3 and 4) using tensorflow that trains on the data and calculates the accuracy score on the test data.

Project Guidelines

Begin by extracting the ipynb file and the data in the same folder. The CNN model (cnn_model_fn) should have the following layers:

- Input layer
- Convolutional layer 1 with 32 filters of kernel size[5,5]
- Pooling layer 1 with pool size[2,2] and stride 2
- Convolutional layer 2 with 64 filters of kernel size[5,5]
- Pooling layer 2 with pool size[2,2] and stride 2
- Dense layer whose output size is fixed in the hyper parameter: fc_size = 32
- Dropout layer with dropout probability 0.4

   Predict the class by doing a softmax on the output of the dropout layers. This should be followed by training and evaluation:

- For the training step, define the loss function and minimize it.
- For the evaluation step, calculate the accuracy.

   Run the program for 100, 200, and 300 iterations, respectively. Follow this by a report on the final accuracy and loss on the evaluation data.

```
from __future__ import absolute_import, division, print_function, unicode_literal
```

Set hyper parameters

- Run the program with three num_steps : 100,200,300

```
import tensorflow as tf
```

```
from tensorflow.keras.models import Sequential
from tensorflow.keras.layers import Dense, Conv2D, Flatten, Dropout, MaxPooling2D
from tensorflow.keras.preprocessing.image import ImageDataGenerator

import os
import numpy as np
import matplotlib.pyplot as plt

train_dir = os.path.join('F:/jupyter/Deep Learning course - Capstone Project/data
validation_dir = os.path.join('F:/jupyter/Deep Learning course - Capstone Project,
```

The model should have the following layers

- input later
- conv layer 1 with 32 filters of kernel size[5,5],
- pooling layer 1 with pool size[2,2] and stride 2
- conv layer 2 with 64 filters of kernel size[5,5],
- pooling layer 2 with pool size[2,2] and stride 2
- dense layer whose output size is fixed in the hyper parameter: fc_size=32
- drop out layer with droput probability 0.4
- predict the class by doing a softmax on the output of the dropout layers

Training

- For training fefine the loss function and minimize it
- For evaluation calculate the accuracy

Reading Material

- For ideas look at tensorflow layers tutorial

```
train_cats_dir = os.path.join(train_dir, 'cats')
train_dogs_dir = os.path.join(train_dir, 'dogs')

validation_cats_dir = os.path.join(validation_dir, 'cats')
validation_dogs_dir = os.path.join(validation_dir, 'dogs')
```

```
total_val = num_cats_val + num_dogs_val
```

```
print('total training cat images:', num_cats_tr)
print('total training dog images:', num_dogs_tr)

print('total validation cat images:', num_cats_val)
print('total validation dog images:', num_dogs_val)
print("--")
print("Total training images:", total_train)
print("Total validation images:", total_val)
```

```
total training cat images: 20
total training dog images: 20
total validation cat images: 10
total validation dog images: 10
--
Total training images: 40
Total validation images: 20
```

```
batch_size = 5
epochs = 15
IMG_HEIGHT = 150
IMG_WIDTH = 150
```

```
train_image_generator = ImageDataGenerator(rescale=1./255) # Generator for our tr
validation_image_generator = ImageDataGenerator(rescale=1./255) # Generator for o
```

```
train_data_gen = train_image_generator.flow_from_directory(batch_size=batch_size,
                                                           directory=train_dir,
                                                           shuffle=True,
                                                           target_size=(IMG_HEIGH
                                                           class_mode='binary')
```

```
Found 40 images belonging to 2 classes.
```

```
val_data_gen = validation_image_generator.flow_from_directory(batch_size=batch_si
                                                              directory=validatio
                                                              target_size=(IMG_HE
                                                              class_mode='binary'
```

```
Found 20 images belonging to 2 classes.
```

```
sample_training_images, _ = next(train_data_gen)
```

```
# This function will plot images in the form of a grid with 1 row and 5 columns
def plotImages(images_arr):
    fig, axes = plt.subplots(1, 5, figsize=(20,20))
    axes = axes.flatten()
    for img, ax in zip( images_arr, axes):
        ax.imshow(img)
        ax.axis('off')
    plt.tight_layout()
```

```
    plt.show()
```

```
plotImages(sample_training_images[:5])
```

```
model_new = Sequential([
    Conv2D(32, (5,5), padding='same', activation='relu',
            input_shape=(IMG_HEIGHT, IMG_WIDTH ,3)),
    MaxPooling2D(),
    Dropout(0.4),
    Conv2D(64, (5,5), padding='same', activation='relu'),
    MaxPooling2D(),
    Dropout(0.4),
    Flatten(),
    Dense(32, activation='relu'),
    Dense(1, activation='sigmoid')
])
```

```
model_new.compile(optimizer='adam',
                loss='binary_crossentropy',
                metrics=['accuracy'])
```

```
model_new.summary()
```

Model: "sequential_3"

Layer (type)	Output Shape	Param #
conv2d_6 (Conv2D)	(None, 150, 150, 32)	2432
max_pooling2d_6 (MaxPooling2	(None, 75, 75, 32)	0
dropout_6 (Dropout)	(None, 75, 75, 32)	0
conv2d_7 (Conv2D)	(None, 75, 75, 64)	51264
max_pooling2d_7 (MaxPooling2	(None, 37, 37, 64)	0
dropout_7 (Dropout)	(None, 37, 37, 64)	0
flatten_3 (Flatten)	(None, 87616)	0
dense_6 (Dense)	(None, 32)	2803744
dense_7 (Dense)	(None, 1)	33

```
Total params: 2,857,473
Trainable params: 2,857,473
Non-trainable params: 0
```

```
history = model_new.fit_generator(
    train_data_gen,
    steps_per_epoch=total_train // batch_size,
    epochs=epochs,
    validation_data=val_data_gen,
    validation_steps=total_val // batch_size
)
```

```
Epoch 1/15
8/8 [==============================] - 5s 649ms/step - loss: 1.2458 - acc: 0.60
00 - val_loss: 0.6930 - val_acc: 0.5000
Epoch 2/15
8/8 [==============================] - 4s 516ms/step - loss: 0.7005 - acc: 0.50
00 - val_loss: 0.6933 - val_acc: 0.5000
Epoch 3/15
8/8 [==============================] - 4s 522ms/step - loss: 0.6919 - acc: 0.50
00 - val_loss: 0.6931 - val_acc: 0.5000
Epoch 4/15
8/8 [==============================] - 4s 523ms/step - loss: 0.6925 - acc: 0.50
00 - val_loss: 0.6925 - val_acc: 0.5000
Epoch 5/15
8/8 [==============================] - 5s 569ms/step - loss: 0.6950 - acc: 0.50
00 - val_loss: 0.6932 - val_acc: 0.5000
Epoch 6/15
8/8 [==============================] - 5s 570ms/step - loss: 0.6891 - acc: 0.55
00 - val_loss: 0.6928 - val_acc: 0.5000
Epoch 7/15
8/8 [==============================] - 5s 569ms/step - loss: 0.7073 - acc: 0.50
00 - val_loss: 0.6932 - val_acc: 0.5000
Epoch 8/15
8/8 [==============================] - 4s 525ms/step - loss: 0.6900 - acc: 0.50
00 - val_loss: 0.6931 - val_acc: 0.5000
Epoch 9/15
8/8 [==============================] - 4s 520ms/step - loss: 0.7233 - acc: 0.52
50 - val_loss: 0.6903 - val_acc: 0.5000
Epoch 10/15
8/8 [==============================] - 4s 523ms/step - loss: 0.6878 - acc: 0.50
00 - val_loss: 0.6904 - val_acc: 0.6500
Epoch 11/15
8/8 [==============================] - 4s 522ms/step - loss: 0.6750 - acc: 0.52
50 - val_loss: 0.6853 - val_acc: 0.5500

Epoch 12/15
8/8 [==============================] - 4s 525ms/step - loss: 0.6702 - acc: 0.62
50 - val_loss: 0.6949 - val_acc: 0.5000
Epoch 13/15
8/8 [==============================] - 4s 519ms/step - loss: 0.6323 - acc: 0.60
00 - val_loss: 0.7244 - val_acc: 0.4500
Epoch 14/15
8/8 [==============================] - 4s 527ms/step - loss: 0.6583 - acc: 0.65
00 - val_loss: 0.6540 - val_acc: 0.8500
Epoch 15/15
8/8 [==============================] - 4s 518ms/step - loss: 0.5336 - acc: 0.87
50 - val_loss: 0.7012 - val_acc: 0.5000
```

```
history_dict = history.history
print(history_dict.keys())
```

```
dict_keys(['loss', 'acc', 'val_loss', 'val_acc'])
```

```
acc = history_dict['acc']
val_acc = history_dict['val_acc']
loss = history_dict['loss']
val_loss = history_dict['val_loss']
```

```
epochs_range = range(epochs)

plt.figure(figsize=(8, 8))
plt.subplot(1, 2, 1)
plt.plot(epochs_range, acc, label='Training Accuracy')
plt.plot(epochs_range, val_acc, label='Validation Accuracy')
plt.legend(loc='lower right')
plt.title('Training and Validation Accuracy')

plt.subplot(1, 2, 2)
plt.plot(epochs_range, loss, label='Training Loss')
plt.plot(epochs_range, val_loss, label='Validation Loss')
plt.legend(loc='upper right')
plt.title('Training and Validation Loss')
plt.show()
```

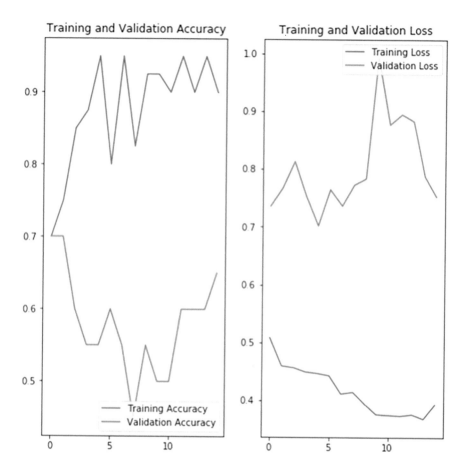

References

Data Set Web References

1. https://www.digitalocean.com/community/tutorials/how-to-build-a-neural-network-to-recognize-handwritten-digits-with-tensorflow
2. https://medium.com/@wainglaminlwin/burmese-handwritten-digit-recognition-with-tensorflow-74021978c509
3. https://towardsdatascience.com/convolutional-neural-networks-357b9b2d75bd
4. https://towardsdatascience.com/simple-introduction-to-convolutional-neural-networks-cdf8d-3077bac
5. https://www.superdatascience.com/blogs/the-ultimate-guide-to-convolutional-neural-networks-cnn
6. https://towardsdatascience.com/a-guide-to-neural-network-layers-with-applications-in-keras-40ccb7ebb57a
7. https://stepupanalytics.com/detailed-introduction-to-recurrent-neural-networks/
8. https://colah.github.io/posts/2015-08-Understanding-LSTMs/
9. https://en.wikipedia.org/wiki/Long_short-term_memory
10. http://www.yaronhadad.com/deep-learning-most-amazing-applications/

References Books

1. Learning TensorFlow [Authors: Tom Hope, Yehezkel S. Resheff & Itay Lieder]
2. Deep Learning Pipeline: Building A Deep Learning Model With TensorFlow [Authors: Hisham El-Amir, Mahmoud Hamdy]
3. TensorFlow for Machine Intelligence_ A Hands-On Introduction to Learning Algorithms [Authors: Sam Abrahams, Danijar Hafner, Erik Erwitt, Ariel Scarpinelli]
4. Python Machine Learning: Machine Learning and Deep Learning with Python, scikit-learn, and TensorFlow [Authors: Sebastian Raschka, Vahid Mirjalili]
5. Python Deep Learning: Exploring deep learning techniques, neural network architectures and GANs with PyTorch, Keras and TensorFlow. [Authors: Ivan Vasilev, Daniel Slater, Gianmario Spacagna, Peter Roelants, Valentino Zocca]

© Springer Nature Switzerland AG 2021
K. B. Prakash, G. R. Kanagachidambaresan (eds.), *Programming with TensorFlow*, EAI/Springer Innovations in Communication and Computing, https://doi.org/10.1007/978-3-030-57077-4

6. Practical Computer Vision Applications Using Deep Learning with CNNs: With Detailed Examples in Python Using TensorFlow and Kivy. [Author: Ahmed Fawzy Gad]

7. Hands-On Machine Learning with Scikit-Learn and TensorFlow: Concepts, Tools, and Techniques to Build Intelligent Systems. [Author: Aurélien Géron]

8. Learn TensorFlow 2.0: Implement Machine Learning And Deep Learning Models With Python. [Authors: Pramod Singh, Avinash Manure]

9. Hands-On Transfer Learning with Python Implement Advanced Deep Learning and Neural Network Models Using TensorFlow and Keras [Authors: Dipanjan Sarkar, Raghav Bali, Tamoghna Ghosh]

10. Hands-On Deep Learning for Images with TensorFlow: Build intelligent computer vision applications using TensorFlow and Keras [Authors: Will Ballard]

11. Deep Learning with Applications Using Python: Chatbots and Face, Object, and Speech Recognition with Tensorflow and Keras. [Authors: Navin Kumar Manaswi]

12. Intelligent mobile projects with TensorFlow : build 10+ artificial intelligence apps using TensorFlow Mobile and Lite for iOS, Android, and Raspberry Pi. [Authors: Tang, Jeff]

13. Intelligent Projects Using Python: 9 real-world AI projects leveraging machine learning and deep learning with TensorFlow and Keras. [Authors: Santanu Pattanayak]

14. Deep Learning in Python: Master Data Science and Machine Learning with Modern Neural Networks written in Python, Theano, and TensorFlow. [Authors: LazyProgrammer]

15. Deep learning quick reference : useful hacks for training and optimizing deep neural networks with TensorFlow and Keras. [Authors: Bernico, Mike]

16. Deep Learning for Computer Vision: Expert techniques to train advanced neural networks using TensorFlow and Keras. [Authors: Rajalingappaa Shanmugamani]

17. Internet of Things for Industry 4.0, EAI, Springer, Editors, G. R. Kanagachidambaresan, R. Anand, E. Balasubramanian and V. Mahima, Springer.

18. Deep Learning with TensorFlow: Explore neural networks with Python [Authors: Giancarlo Zaccone, Md. Rezaul Karim, Ahmed Menshawy]

19. TensorFlow 1.x Deep Learning Cookbook: Over 90 unique recipes to solve artificial-intelligence driven problems with Python. [Authors: Antonio Gulli, Amita Kapoor]

20. Hands-on unsupervised learning with Python : implement machine learning and deep learning models using Scikit-Learn, TensorFlow, and more [Authors: Bonaccorso, Giuseppe]

21. Mastering TensorFlow 1.x: Advanced machine learning and deep learning concepts using TensorFlow 1.x and Keras. [Author: Armando Fandango]

22. Practical Deep Learning for Cloud, Mobile, and Edge: Real-World AI & Computer-Vision Projects Using Python, Keras & TensorFlow [Authors: Anirudh Koul, Siddha Ganju, Meher Kasam]

Article References

23. Agrawal A, Roy K (2019) Mimicking leaky-integrate-fire spiking neuron using automotion of domain walls for energy-efficient brain-inspired computing. IEEE Trans Magn 55(1):1–7

24. Akinaga H, Shima H (2010) Resistive random access memory (reram) based on metal oxides. Proc IEEE 98(12):2237–2251

25. Amit DJ, Amit DJ (1992) Modeling brain function: the world of attractor neural networks Cambridge University Press, Cambridge

26. Bourzac K (2017) Has intel created a universal memory technology?[news]. IEEE Spectr 54(5):9–10

27. Goodfellow I, Bengio Y, Courville A, Bengio Y (2016) Deep learning, vol 1. MIT Press, Cambridge

28. Grandison T, Sloman M (2000) A survey of trust in internet applications. IEEE Commun Surv Tutor 3(4):2–16

29. Guo X, Ipek E, Soyata T (2010) Resistive computation: avoiding the power wall with low-leakage, STT-MRAM based computing. In: ACM SIGARCH computer architecture news, vol 38. ACM, pp 371–382

30. Hurst S (1969) An introduction to threshold logic: a survey of present theory and practice. Radio Electron Eng 37(6):339–351

31. Jeong H, Shi L (2018) Memristor devices for neural networks. J Phys D: Appl Phys 52(2):023003

32. Krestinskaya O, Dolzhikova I, James AP (2018) Hierarchical temporal memory using memristor networks: a survey. IEEE Trans Emerg Top Comput Intell 2(5):380–395. https://doi.org/10.1109/TETCI.2018.2838124

33. Krestinskaya O, James AP, Chua LO (2019) Neuro-memristive circuits for edge computing: a review. IEEE Trans Neural Networks Learn Syst (https://doi.org/10.1109/TNNLS.2019.2899262). arXiv:1807.00962

34. Krestinskaya O, Salama KN, James AP (2018) Learning in memristive neural network architectures using analog backpropagation circuits. IEEE Trans Circuits Syst I: Regul Pap 1–14. https://doi.org/10.1109/TCSI.2018.2866510

35. Krestinskaya O, Bakambekova A, James AP (2019) Amsnet: analog memristive system architecture for mean-pooling with dropout convolutional neural network. In: IEEE international-conference on artificial intelligence circuits and systems

36. Li Y, Wang Z, Midya R, Xia Q, Yang JJ (2018) Review of memristor devices in neuromorphic computing: materials sciences and device challenges. J Phys D: Appl Phys 51(50):503002

37. Liao Q, Poggio T (2016) Bridging the gaps between residual learning, recurrent neural networks and visual cortex. arXiv:1604.03640

38. Lippmann R (1987) An introduction to computing with neural nets. IEEE ASSP Mag 4(2):4–22

39. Ma J, Tang J (2017) A review for dynamics in neuron and neuronal network. Nonlinear Dyn 89(3):1569–1578

40. Maan AK, Jayadevi DA, James AP (2017) A survey of memristive threshold logic circuits. IEEE Trans Neural Netw Learn Syst 28(8):1734–1746

41. McCulloch WS, Pitts W (1943) A logical calculus of the ideas immanent in nervous activity. Bull Math Biophys 5(4):115–133

42. Medsker L, Jain LC (1999) Recurrent neural networks: design and applications. CRC Press,

43. Boca Raton Mhaskar H, Liao Q, Poggio T (2016) Learning functions: when is deep better than shallow. arXiv:1603.00988

44. Nili H, Adam GC, Hoskins B, Prezioso M, Kim J, Mahmoodi MR, Bayat FM, Kavehei O

45. Strukov DB (2018) Hardware-intrinsic security primitives enabled by analogue state and non-linear conductance variations in integrated memristors. Nat Electron 1(3):197

46. Raghu M, Poole B, Kleinberg J, Ganguli S, Dickstein JS (2017) On the expressive power of deep neural networks. In: Proceedings of the 34th international conference on machine learning-volume 70. pp. 2847–2854. https://www.JMLR.org

47. Schlkopf B, Smola AJ, Bach F (2018) Learning with kernels: support vector machines, regularization, optimization, and beyond

48. Ren S, He K, Girshick RB, Sun J (2017) Faster r-cnn: towards real-time object detection with region proposal networks. IEEE Trans Pattern Anal Mach Intell 39(6):1137–1149

49. M. Abadi, A. Agarwal, P. Barham, E. Brevdo, Z. Chen, C. Citro, G. S. Corrado, A. Davis, J. Dean, M. Devin, S. Ghemawat, I. Goodfellow, A. Harp, G. Irving, M. Isard, Y. Jia, R. Jozefowicz, L. Kaiser, M. Kudlur, J. Levenberg, D. Mane, R. Monga, S. Moore, D. Murray, C. Olah, M. Schuster, J. Shlens, B. Steiner, I. Sutskever, K. Talwar, P. Tucker, V. Vanhoucke, V. Vasudevan, F. Viegas, O. Vinyals, P. Warden, M. Wat- ́ tenberg, M. Wicke, Y. Yu, and X. Zheng. TensorFlow: Large-scale machine learning on heterogeneous systems, 2015. Software available from tensorflow.org

50. D. Archambault, T. Munzner, and D. Auber. Grouseflocks: Steerable exploration of graph hierarchy space. Visualization and Computer Graphics, IEEE Transactions on, 14(4):900–913, 2008.

51. J. Abello, F. Van Ham, and N. Krishnan. ASK-Graphview: A large scale graph visualization sytem. Visualization and Computer Graphics, IEEE Transactions on, 12(5):669–676, 2006.

52. C.C. Paige, M. Saunders, LSQR: an algorithm for sparse linear equations and sparse leas squares, ACM Transactions on Mathematical Software 8 (1982) 43–71.

53. M. Abramowitz, I.A. Stegun, Handbook of Mathematical Functions, in: National Bureau of Standards, Series, vol. #55, Dover Publications, USA, 1964.

54. R.R. Hocking, Methods and Applications of Linear Models, in: Wiley Series in Probability and Statistics, Wiley-Interscience, New York, 1996.

55. J.R. Magnus, H. Neudecker, Matrix Differential Calculus with Applications in Statistic and Econometrics, revised ed., in: Wiley Series in Probability and Statistics, John Wiley & Sons, Chichester, UK, 1999.

56. G.A.F. Seber, The Linear Hypothesis: A General Theory, in: Griffin's Statistical Monographs and Courses, Charles Griffin and Company Limited, London, 1966.

57. S.F. Ashby, M.J. Holst, T.A. Manteuffel, P.E. Saylor, The role of the inner product in stopping criteria for conjugate gradient iterations, BIT 41 (1) (2001) 26–52.

58. O. Axelsson, I. Kaporin, Error norm estimation and stopping criteria in preconditioned conjugate gradient iterations, Journal of Numerical Linear Algebra with Applications 8 (2001) 265–286

59. Baron, R.A., & Ensley, M.D. 2005. Opportunity recognition as the detection of meaningful patterns: Evidence from the prototypes of novice and experienced entrepreneurs. Manuscript under review

60. Ren S, He K, Girshick RB, Sun J (2017) Faster r-cnn: towards real-time object detection with region proposal networks. IEEE Trans Pattern Anal Mach Intell 39(6):1137–1149

61. Baron, R.A., & Ensley, M.D. 2005. Opportunity recognition as the detection of meaningful patterns: Evidence from the prototypes of novice and experienced entrepreneurs. Manuscript under review

62. Smola, A. and Sch¨olkopf, B. On a kernel-based method for pattern recognition, regression, approximation and operator inversion. Algorithmica (to appear), 1998.

63. Osuna, E. and Girosi. F. Reducing the run-time complexity of support vector machines. In International Conference on Pattern Recognition (submitted), 1998

64. Osuna, E., Freund, R. and Girosi, F. Training support vector machines: an application to face detection. In IEEE Conference on Computer Vision and Pattern Recognition, pages 130 – 136, 1997.

65. Devroye, L., Gyorfi, L. and Lugosi, G. A Probabilistic Theory of Pattern Recognition. Springer Verlag, Applications of Mathematics Vol. 31, 1996.

66. Aizerman, M.A., Braverman, E.M. and Rozoner, L.I. Theoretical foundations of the potential function method in pattern recognition learning. Automation and Remote Control, 25:821–837, 1964

Index

© Springer Nature Switzerland AG 2021
K. B. Prakash, G. R. Kanagachidambaresan (eds.), *Programming with
TensorFlow*, EAI/Springer Innovations in Communication and Computing,
https://doi.org/10.1007/978-3-030-57077-4